Sebastian Gieselmann

Development of a robotic system to design and evaluate Dynamic Background Cues

disserta Verlag

Gieselmann, Sebastian: Development of a robotic system to design and evaluate
Dynamic Background Cues, Hamburg, disserta Verlag, 2013

ISBN: 978-3-95425-246-6
Druck: disserta Verlag, ein Imprint der Diplomica® Verlag GmbH, Hamburg, 2013
Coverbild: Sebastian Gieselmann, Herford, 2010

Bibliografische Information der Deutschen Nationalbibliothek
Die Deutsche Nationalbibliothek verzeichnet diese Publikation in der Deutschen
Nationalbibliografie; detaillierte bibliografische Daten sind im Internet über
http://dnb.d-nb.de abrufbar.

Die digitale Ausgabe (eBook-Ausgabe) dieses Titels trägt die ISBN 978-3-95425-247-3
und kann über den Handel oder den Verlag bezogen werden.

Zugl.: Bielefeld, Universität Bielefeld, Dissertation, 2013

Dieses Werk ist urheberrechtlich geschützt. Die dadurch begründeten Rechte, insbesondere die der Übersetzung, des Nachdrucks, des Vortrags, der Entnahme von Abbildungen und Tabellen, der Funksendung, der Mikroverfilmung oder der Vervielfältigung auf anderen Wegen und der Speicherung in Datenverarbeitungsanlagen, bleiben, auch bei nur auszugsweiser Verwertung, vorbehalten. Eine Vervielfältigung dieses Werkes oder von Teilen dieses Werkes ist auch im Einzelfall nur in den Grenzen der gesetzlichen Bestimmungen des Urheberrechtsgesetzes der Bundesrepublik Deutschland in der jeweils geltenden Fassung zulässig. Sie ist grundsätzlich vergütungspflichtig. Zuwiderhandlungen unterliegen den Strafbestimmungen des Urheberrechtes.

Die Wiedergabe von Gebrauchsnamen, Handelsnamen, Warenbezeichnungen usw. in diesem Werk berechtigt auch ohne besondere Kennzeichnung nicht zu der Annahme, dass solche Namen im Sinne der Warenzeichen- und Markenschutz-Gesetzgebung als frei zu betrachten wären und daher von jedermann benutzt werden dürften.

Die Informationen in diesem Werk wurden mit Sorgfalt erarbeitet. Dennoch können Fehler nicht vollständig ausgeschlossen werden und der Verlag, die Autoren oder Übersetzer übernehmen keine juristische Verantwortung oder irgendeine Haftung für evtl. verbliebene fehlerhafte Angaben und deren Folgen.

© disserta Verlag, ein Imprint der Diplomica Verlag GmbH
http://www.disserta-verlag.de, Hamburg 2013
Hergestellt in Deutschland

Acknowledgements

I would like to thank everyone who supported me during my research in advice as well as in time. Furthermore I acknowledge the CorLab for funding me and providing me the hardware I needed to conclude my work. Of cause I thank my supervisor Britta Wrede for supporting me all the time, and my committee for reviewing my thesis.

Last but not least I would like to express my sincere thanks to my whole family for giving me the opportunity and the sturdiness to finish this thesis.

Abstract

Robots are becoming more and more important in modern society. Often acceptance of especially human-like robots is not very high because people are afraid of this new technology. Especially in the western world, this trend can be observed in literature and movies where robots became dangerous. On the other hand, many researchers are working on human robots e.g. for elderly care or as household helpers. But what is the benefit of those robots if only a few people want to use them?

The question that is addressed in this thesis is how subtle movements can influence the way people perceive robots, and how robots can become friendlier and more acceptable in the eyes of the user. The proposed approach addresses social behaviors such as idle movements, blinking, and lip-movements. Despite not having any functional meaning within an interaction, these cues influence a persons perception of the robot as well as their physiology. This thesis is the start of a systematic investigation of the influence of these behaviors, and the ways they can be used to improve Human- Robot Interaction. To achieve this goal, a robotic system was developed to makes design and use of social behavior easier. Furthermore, studies have been carried out to examine the perceived familiarity, arousal level and task performance of the human participants using questionnaires and biophysiological measures.

Contents

Contents	7
List of Figures	10
1 Introduction	**13**
1.1 Overview	18
2 Theoretical Background	**19**
2.1 Human Communication	21
2.2 Perception of Robots	26
2.2.1 Anthropomorphism	27
2.2.2 The Mindlessness Approach	31
2.2.3 Form and Function	32
2.2.4 Conclusion	33
2.3 Arousal and Task Performance	34
2.3.1 Definitions	34
2.3.2 Humans and Robots in Cooperative tasks	38
2.4 Measurement of Stress	40
3 Definitions and Hypotheses	**45**
3.1 Typology of Signals and Cues	45
3.1.1 The human designer	48
3.1.2 Dimension 1: Source	50
3.1.3 Dimension 2: Type	51
3.1.4 Dimension 3: Movement Type	51
3.1.5 Dimension 4: Perceptual Type	53
3.2 Definition of Dynamic Background Cues	53
3.3 Hypotheses	56
4 Review of relevant Social Robots	**59**
4.1 Criteria for Social Robots	59

CONTENTS

4.2	Kismet/Leonardo	63
4.3	Nao	66
4.4	Geminoid	67
4.5	Barthoc Jr./Flobi	69
4.6	iCub	70
4.7	Miscellaneous Systems	72
4.8	Implications	74

5 Proof of Concept   79
5.1	Social behavior	80
5.2	Study design	80
5.3	Measures	84
5.4	Results	86
5.5	Discussion	88

6 Dynamic Background Cue Pattern   91
6.1	Blinking	93
6.2	Natural Movements	95
6.3	Idle Movements	99
6.4	Gaze	102
	6.4.1 Online Study	106
6.5	Lip-Movement	114
	6.5.1 Online Study	118

7 System Description   127
7.1	Overview	128
7.2	Systemstructure	130
	7.2.1 Test Cases	135
7.3	PiRA-XML	138
	7.3.1 The structure of PiRA commands	141
7.4	Memory Layer	146
7.5	Hardware Abstraction Layer: Arbiter	149
7.6	Hardware Abstraction Layer: Command server	152
7.7	Preprocess	157
7.8	Social Layer	159
7.9	Exchangeability	162
7.10	PiRA-Editor	162
	7.10.1 Program description	163
	7.10.2 Usability Study	167

8 Studies **173**
8.1 Whole Body Study 174
8.1.1 Setup 174
8.1.2 Measurements 176
8.1.3 Results 177
8.1.4 Discussion 180
8.2 Interaction Study 183
8.2.1 Setup 183
8.2.2 Measurements 189
8.2.3 Results 195
8.2.4 Discussion 201

9 Conclusions **207**

A Questionnaires **213**
A.1 Proof of concept 213
A.2 Online Studies 215
A.3 Usability Studies 217
A.4 Whole Body Study 221
A.5 Interaction Study 222

B Tables **225**
B.1 Proof of concept 225
B.2 Online Study: Gazing 226
B.3 Phoneme/Viseme mapping 227
B.4 Online Study: Lips 228
B.5 Online Study: Whole Body 228
B.6 Interaction Study: iCub conversation 228
B.7 Interaction Study: Statistics 230

C Images **235**

D UML Diagrams **239**

E Picture Credits **243**

List of Figures

1.1	Examples of social robots	15
2.1	Relation of signs	22
2.2	The Uncanny Valley	27
2.3	Example of objects that are anthropomorphised	28
2.4	Determinations of work performance	35
2.5	The Yerkes-Dodson Law	37
2.6	EKG and ESR Sensors	41
2.7	Sample ESR plot	42
3.1	Typology of human-like and artificial signals and cues for robotics	47
3.2	Examples of artificial signals	51
3.3	Importance of functional and background signs for different movements	54
4.1	The social Layer	61
4.2	Kismet and Leonardo	64
4.3	Humanoid robot Nao	66
4.4	Humanoid robot Geminoid	67
4.5	Humanoid robots Barthoc Jr and Flobi	69
4.6	Open source humanoid robot iCub	71
4.7	Miscellaneous social robots	72
5.1	Participant performing six tutoring tasks	82
5.2	Study setup	84
6.1	Different stages of iCub performing a blink	94
6.2	Pointing gesture	96
6.3	Transferring human motion to a robot using video data	98
6.4	Social and non-social posture of iCub	100
6.5	Breathing animation for iCub.	102
6.6	Different stages of iCub performing the gaze DBC	105
6.7	Timing of mutual gaze during the online study	107

List of Figures

6.8	Possible visemes of iCub	116
7.1	Different strategies to communicate with a robot	127
7.2	Comparing Social Layer Architecture (SLA) and direct communication	128
7.3	Block diagram illustrating the Social Layer Architecture	133
7.4	Execution time needed to execute a command in the SLA	136
7.5	Three different animation strategies	140
7.6	PiRA data structure as UML Class Diagram	142
7.7	Joint movements depend on acceleration	154
7.8	Execution time needed to generate a posture during an animation.	155
7.9	Graphical User Interface of the PiRA-Editor.	164
7.10	Example for a discrete joint	165
7.11	Timeline window from Choregraphe	166
8.1	Schematic diagram of the setup used in the interaction study	184
8.2	Possible solutions for the three building tasks.	185
8.3	Pictures of the setup used in the interaction study	186
8.4	Different gestures produced during the interaction Study.	188
8.5	Example computation of the score needed to compare the ESR data	194
A.1	Questionnaire: Proof of concept 1	213
A.2	Questionnaire: Proof of concept 3	213
A.3	Questionnaire: Proof of concept 2	214
A.4	Instructions for the online studies	215
A.5	Video page from the online studies	215
A.6	Questionnaire for the gaze online study	216
A.7	Questionnaire for the lip online study	216
A.8	Exercise sheet for the PiRA-Editor	217
A.9	Exercise sheet for Choregraphe	218
A.10	Questionnair used to rate the usability of the editors	219
A.11	Questionnair for personal data	220
A.12	First part of the questionnaire used to rate the robot	221
A.13	Second part of the questionnaire used to rate the robot	221
A.14	Goodspeed Questionnaire created by Bartneck et al.	222
C.1	Corresponding visemes and lip postures	235
C.2	Idle movements: Looking around as performed by iCub	236
C.3	Idle movements: A stretchung motion performed by iCub	237
C.4	Pointing gesture	238

List of Figures

D.1	Sequence Diagram: External command	239
D.2	Sequence Diagram: Gaze DBC	240
D.3	Sequence Diagram: Speech command	241
D.4	Class Diagram: `CommandServer`	242

Chapter 1

Introduction

When predicting the technological development of assistive systems in everyday environments, it is save to assume that the future home will contain more electronic devices than it does today. As technology becomes smaller, more efficient and more diverse, it carries with it the promise that it might help people in unprecedented ways, perhaps even fulfilling all their physical needs. At least if you believe the reality suggested by science fiction such as Star Trek.

We already have a great deal of technology and even robots in our homes. For example vacuum cleaner robots are used to clean floors fully autonomously. People are becoming more accustomed to robots. Even the elderly are learning to interact naturally with the seal robot Paro [154]. However problems may emerge as robots become more and more human-like. At the moment most robots available have a technical appearance. But as robots are developed to address more complex tasks i.e. for serving as social interfaces or housekeeping, it becomes more reasonable to design human-like robots. This shape has the advantage that it is natural for people to interact with other people, and that it fits easily in households designed for human use. But it is exactly this human-like shape that often evokes fear and discomfort. When the robot becomes too human-like or too powerful and intelligent, especially people in the western world are often scared when they are confronted with such robots [81].

This vision of what robots could be or do in the future can be seen in literature or movies where authors express their fear that could robots become dangerous for humans, as seen in the "Terminator" movies or stories like I,Robot by Isaac Asimov. This leads people to have an incorrect view of what "real" humanoid robots are like. Elderly people in particular express their concerns when saying "They will take our jobs" or asking "when will robots take over the world? "

1. INTRODUCTION

In order to reduce these fears it has become important to convince people that robots are friendly and that it is pleasant to interact with them. In practice it is difficult to achieve this goal. Often it is not possible for the hardware to appear social. Alternately, the behavior of a robot can generate misunderstandings that make interaction uncomfortable for the user. This is why researchers from all over the world are working to create robots that are accepted by the human and capable of interacting naturally with them.

Designing social human-like robots is a challenging task where many factors have to be considered. The challenges are not only technical but also tightly connected to social science, making social robotics a very interdisciplinary field. At first sight, the analysis of features that affects the perception of social robots can be divided into two large fields, comprising the static and dynamic features of the robot. As soon as a robot presents at least some social features, either static or dynamic, the robot can be called social. Thats why, a robot can be classified as a social robot if it influences humans socially, regardless of whether it can exhibit these features to a lesser or greater degree.

Static features in this context refer to the outer appearance of the robot. According to Hegel [69] it should be designed with respect to the functionality of the robots and should be socially acceptable. The design is the basis for social behavior and reveals necessary behaviors or limitations. For example features such as ears give users the illusion that the robot can understand spoken language or acoustic signals. As such, a robot that has ears should be able to understand spoken language. In contrast, a robot without hands is not able to do hand gestures. Duffy states that "there should be a strong correlation between the robot's capabilities and its form and vice versa" [43]. Much work has already been done in the area of social robot design (Hegel, [69]). The design of a social robot need not necessarily be human-like (e.g.: Geminoid DK in Figure 1.1 (right)) although this appearance is most familiar to humans. The robot Kismet is a good example of a robot designed with a more artificial appearance (e.g.: Kismet in Figure 1.1 (middle)).

It is far more important to design robots in such a way that the user does not feel intimidated [51] of afraid [43]. A robot with a height of 2 meter may be frightening or a robot that looks very intelligent may make the user feel stupid. Many social robots that are actually used are designed to look like little children so that they appear weak and less intelligent. This design suits the actual capabilities of robots and makes them appear more

Figure 1.1: Example social robots: iCub[106]; Kismet[21]; Geminoid DK[29](from left to right)

cute and friendly (e.g.: iCub in Figure 1.1 (left)). Beyond these, other approaches have been developed for the design of social robots.
Dynamic features encompass movement characteristics such as gestures, idle movements, speech or manipulation of objects. When these features are socially motivated they are called social behavior
The goal of this thesis is to examine the way in which dynamic features in robots alter human perception of the robot, and to investigate whether this behavior can be used to improve performance in cooperative tasks. The social behaviors we focus on are called Dynamic Background Cues (DBC), and are equivalent to human involuntary dynamical signs such as blinking, breathing and lip-movements. Furthermore the background in DBC means that these cues have no direct influence on the task of the robot. Often social behavior as reported in literature, consists of a functional behavior that directly influences the task, for example gazing, that then reveals information about the attention of the robot towards a desired object (i.e. the position). Functional in this context refers to how the movement provides task relevant information as opposed to background information. The movement can be a pointing gesture, a verbal explanation or the manipulation of objects.
The main targets of this thesis are the background features. For example speech is a dynamic functional behavior which provides a great deal of task relevant information. However, lip movement has no task relevant information and can therefore be classified as DBC. The advantage of using DBCs is that they can be added to an existing functional robotic system to make it more social and lively without influencing the other functions of the robot. The hypothesis is that a robot that uses DBCs appears more lively and familiar to humans. Furthermore, if the robot appears more familiar and lively this can improve task performance since humans will be

1. INTRODUCTION

less disturbed by distracting features of the robot.

It has been found that social cues make robots more likeable and improve task performance [20], but questions remain as to the particular effects of single social cues. The literature does not provide a structured study of single cues,as often these cues are examined in combination with other cues or they are combined with functional cues that provide additional information ([20][114]). If, for example, the robot uses eye gaze as a cue and looks in the direction of the task goal it should be obvious that this social behavior improves task performance when the task for the human is to guess the desired object. This gazing gesture provides functional information about the task and therefore is a functional cue. Evidence of the effectiveness of functional gazing to improve task performance was given for example by Pitsch et al. [121].The authors evaluated task relevant gazing in a task where infants indicated to their parents that they had understood the target task by looking to the goal position very early. The parents were able to improve their tutoring depending on the functional gaze of the children. Because of this display of understanding the parent was able to improve task performance. In parallel A. Vollmer [152] showed that this behavior can also be applied to robots that can help the tutor by using gazing behavior similar to infants.

To realize these investigations simple DBCs were designed, and then evaluation whether interaction was positively affected were carried out. Given that human communication is amazing complex, a selection of only five non-functional social cues, ranging from simple low level features to more complex ones, were chosen as a starting point. The following DBCs where chosen because they are the most visible cues within HRI, and because of their precedent in the literature i.e. Breazeal [20] or in the Geminoid DK [29]. Furthermore similar cues are often chosen by other researchers to make a robot more social:

- **Blinking:** Simulation of human blinking behavior
- **Idle movements:** Keeping the robot in continuous motion so that it never appears stiff.
- **Natural movements:** Adding human-like movement pattern onto functional movements
- **Gaze:** Random gaze shifts during face to face communication
- **Lip-movement:** Moving of lips synchronously to speech

Most of these cues are designed as parameterized movement patterns so that they can be ported to different robots and are not limited to the iCub used in this thesis. This allows to transport the effects that are investigated in this thesis into ones own system. Further, the parameters can be manipulated to generate different character traits for the robot. DBCs which are not on this list will be evaluated in future research.

The open question is what the influences of specific social cues are and how they can be used to improve Human-Robot Interaction (HRI). As such, subjective as well as objective measurements were used. Subjective measures took the form of questionnaires used to evaluate emotions and character traits ascribed to the robot. The objective measures was arousal during interaction as determined by the biophysiological features Electrical Skin Resistance (ESR) and heart rate (ECG). Since arousal is connected to task performance this value can be used to get inferences of robot behavior to task performance.

To evaluate how DBCs influence task performance, arousal, perceived character traits or perceived familiarity, a structured investigation was carried out consisting of online studies, as well as an interaction study with a real robot iCub. During the first part of these studies, single cues were tested to figure out what influences they had. In contrast to many other research parameterized pattern were used to define and configure the dynamics of each DBC. This made it possible to test the effects of different parameters and to port the pattern to similar robots. The last part of the studies focused on the influences of the participant's stress level while interacting with a robot. There were two conditions, one using DBCs and one without. To carry out the investigations, a flexible robotic system was created to make it possible to create new DBCs, and use them on the desired robot. The system was designed to function as a social layer that encapsulates the robot and produces an intrinsic social behavior by providing high-level interfaces in parallel that in turn allow the attachment of any functional system. The purpose of this layer is to produce a social robot that can be used by the functional system the same as any other non-social robot without regard to its social features. For example if the functional system sends a speech command to the robot, corresponding lip-movements would be generated automatically and displayed in synchrony to speech. Further, a tool is implemented and evaluated to facilitate the design of social behavior or body movements. This system was used in all studies to produce stimuli or to steer the HRI during the interaction study.

1. INTRODUCTION

1.1 Overview

This thesis is structured as follows. In chapter 2 all necessary definitions and theoretical background such as human communication and the influence of stress onto performance are given. Some of these definitions are used as a basis for new definitions such as the typology of signals and cues for robots given in chapter 3. In chapter 4 different types of social robots are presented and compared. In chapter 5 the first study is presented which functions as a proof of concept to test if DBCs in general can influence the user. chapter 6 provides the background and the design of each DBC and describes the online studies used to evaluate two of them. The construction of the robotic system is presented in chapter 7, comprising the technical part of this thesis. Finally the two main studies are presented in chapter 8 followed by the conclusion in chapter 9.

Chapter 2

Theoretical Background

In this chapter all necessary information important for understanding the structure of Human-Robot Interaction (HRI) and determinants of task performance that appear when interacting with a social robot will be given. A social robot is a robot that has technical as well as social aspects [69]. Social aspects include the social appearance of the robot as well as its capability for social behavior. The junction between social and non-social robots is fluid. Is an industrial robot arm a social robot because it uses social conventions or can move in a social manner? Is a robot that looks like a person social despite not being able to move autonomously and is instead teleoperated? Often it is difficult to distinguish clearly between both classes. In this thesis a robot is called a social robot as soon as it contains at least some social features, even when these features encompass only appearance or function. A more detailed definition of social robots is given in section 4.1. This project will focused on human-like social robots. This section starts with the general theoretical background for communication and anthropomorphism. Following, task performance in cooperative tasks will be explained. The chapter closes with background information focusing on biophysiological measurements.

To establish a successful interaction it is important to minimize the amount of misunderstandings. In Human-Human Interaction (HHI) small mistakes can create miscommunication between human speakers [34]. For example in different countries [111] the same hand gesture can have different meanings leading to serious misleading signs. As such, signs are small communication units that structure communication. Not only the intended information are important, often the unintended ones are even more important. Humans communicate all the time and mostly by using unintended information about their internal state like emotions, physical

2. THEORETICAL BACKGROUND

condition and so on. If someone is sad or weak because of an illness other humans can see the signs that reveal their illness. Humans are very good at interpreting those signs because they practice their whole lives. In HRI the interaction might follow the same pattern as in HHI because humans anthropomorphise robots and other non-human objects [49]. This is why it is essential to better understand what signs in HHI are interpreted in which way as they could play an important role in HRI, too. Humans expect signs and conventions within an interaction that have to be fulfilled, in HHI as well as in HRI, to minimize misunderstandings and errors during interaction. How human communication is structured is presented in section 2.1. How this can be applied in HRI is presented in section 2.2.

In addition, a lower error rate is not the only measure of good interaction. Performance is another aspect to be considered. When two entities conduct a cooperative task it could be seen as a greater success if the task is done in a shorter time or with greater accuracy. Therefore there is a connection between the success of an interaction and performance. Obviously an interaction with a lower rate of misunderstandings can have higher efficiency but another effect is also very interesting in the context of social robotics. A person's arousal can influence their performance. This arousal is represented by the arousal level. A person that is stressed or bored has a lower performance than a person with a moderate arousal level. The arousal is influenced by many different stressors such as task load, noise or social pressure [102]. Stressors are defined as external sources of stress and thus robots as interaction partners are also stressors. Influence on task performance occur depending on the robots behavior as it is typical for stressors. Subsection 2.3 provides background on arousal and its influence on the interaction.

Communication between two entities can be seen as a situation where both influence each other by exchanging signs. According to Maynard-Smith [104] each sign produced evokes a reaction by the receiver that can then be interpreted by others. Watzlawick says in the third axiom of communication [160] that each individuum sees its own behavior as a reaction to the behavior of others. This shows that communication is a complex system where mutual influence directs the whole interaction. This makes it even more important for robots to react in a socially adequate manner so as to not interrupt the interaction loop. Lohse et al. [94] found that mistakes within HRI that lead to disconfirmations of expectations impair fluent interaction because humans have to cope with this situation and have to find another way to adapt to the robot. Putting all this together forms the hypothesis that providing adequate social behavior can improve the HRI and performance. This hypothesis has already been formulated by others. For example Breazeal conducted two studies where the effects of emotions [15] or the effect of nonverbal communication [20] were examined to see how they improved interaction and performance. In addition, Mutlu et al. [114][113] showed that if a robot unintendedly gazes at its chosen object it increases the performance when playing a guessing game. In contrast to previous studies this thesis focuses on social cues that provide no additional task relevant information. Here cues such as breathing and blinking are used. A complete definition of these Dynamic Background Cues (DBC) will be given in chapter 3.2.

It should also be mentioned that many of the following theories can also be applied to objects such as pets, toys, animal-like robots and so forth. In this work the focus will remain strictly on humanoid robots if not stated otherwise.

2.1 Human Communication

As mentioned before the first step is to figure out how communication in HHI is defined so as to adapt it to HRI. To this end, definitions from fields like semiotics and biology are used. The first and most comprehensive definition is for signs. A sign is defined as everything that transports a meaning. A good summary of signs is given in the handbook of semiotics by Nöth [117]. Nöth says:

> "The word *sign* is ambiguous. It has either the broader sense of a semiotic entity which unites a sign vehicle with a meaning, or it has the narrower sense of a sign vehicle only." ([117] p.79)

In other words everything that can transport a meaning is a sign. A pointing gesture is more then just moving the finger in a certain direction. An observer interprets this for example as an hint to go into a certain direction. Because of this interpretation this gesture attains a meaning and becomes a sign. Sign vehicles do not transport the meaning by itself. It is always the observer that interprets the sign vehicle and ascribes meaning. This is why the same sign can have different meanings depending on the observer, as Watzlawick's states in his third axiom of communication [160], which may lead to misunderstandings.

2. THEORETICAL BACKGROUND

Communication by sending signs was extensively researched within the signaling theory originally given by Spence [147] (see Connelly et al. for a review[36]). Signaling theory first focused on the job market and was then extended to general communication. It describes the behavior of two parties that have access to different information. One party is the signaler and chooses how information is communicated. The other party, as an interpreter, has to choose how to interpret the signal. Another key aspect of signaling theory is the reliability of signals. It says that signals should have a certain quality or must be difficult to fake to convince the receiver. For example a college degree signals a job seeker's intelligence and ability to a prospective employer. Additionally a college degree is difficult to fake.

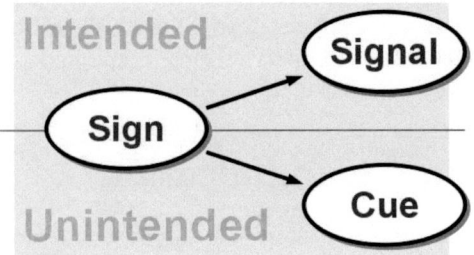

Figure 2.1: Relation of signs, signals and cues

Maynard-Smith is a biologist and adapted signaling theory to animal behavior. He produced definitions that are relevant to HRI. This thesis uses definitions by Maynard-Smith and Harper [104]. Both authors divided signs into signals and cues (see Figure 2.1) to describe the communication pattern between animals. Signals are ...

> "... any act or structure which alters the behaviour of other organisms, which evolved because of that effect, and which is effective because the receiver's response has also evolved." ([104] p.3)

This means signals are behaviors that are produced intentionally to influence the recipient. The signal may be a simple gesture or manipulation of objects. In addition it should be mentioned that the signals defined by Maynard-Smith are different from those used in signaling theory. The signals used in signaling theory can be compared with the signs used in this definition. On the other hand cues are defined as ...

> "... any feature of the world, animate or inanimate, that can be used by an animal as a guide to future actions." ([104] p.3)

These are unintentionally given signs present in the entity's appearance or produced through unintended movements such as breathing or blinking.

For clarification the differences between both signs will be shown in the following two examples. Gazelles show a strange behavior when they spot an enemy like a lion. They start jumping on the spot instead of running away. At first sight this behavior makes no sense because it wastes precious energy necessary for flight. But this behavior is a signal that evolved over a long time. The gazelle signals that it is very strong and healthy by jumping as high as possible. The lion on the other hand knows that it is a waste of energy for him to pursue a strong gazelle if there are other less strong gazelles available. The healthy gazelle is saved by jumping very high and the lion gets a hint about which victim is most worth of pursuit. This phenomena of wasting energy or resources to produce reliable signals to gain a benefit can be observed in many different species. Zahavi [166] called this the handicap principle. A good example of a cue can be found in the amazing tail of male peacocks whose main purpose is to attract females. The bigger and more colorful it is, the more attractive the male peacock is to other females. The trade-off is that this peacock also attracts more predators and greater difficulty running away. Yet this is exactly the reason why those peacocks are more attractive to females: they show that they are healthy and fast enough to flee from enemies even though they have a large and colorful tail. This tail is a reliable sign that shows that this peacock is a good catch for female peacocks. It is also a cue because the peacock shows its tail unintentionally, as such it belongs to the appearance of this animal. In contrast to the gazelles that can decide when to show their jumping signal, peacocks unintentionally show their tails. This is the most important difference between signals and cues.

As the goal is to investigate the parallelism of human communication and HRI, the definitions of animal signs have to be adopted to HHI. Maynard-Smith himself explained in the last chapter of his book [104] that their definition does also fit for humans. In addition Cronk [39] postulated that animal signals and cues are applicable to human communication, al-

2. THEORETICAL BACKGROUND

though he stipulated that additions were necessary to fully adapt the theory of Maynard-Smith to humans. For example humans are very good in faking signals to gain a greater benefit. For now it can be assumed that the concept of signals and cues are applicable for humans. The topic of deception will be discussed later.

Human signals and cues look nearly the same as animal signs. For example some rich people love to show their wealth by spending more money than necessary. They buy expensive cars or jewelry to show everyone their luxurious life. This is a signal as per the given definition. These people willingly decide to show all this luxury in public and the public knows that these items are very expensive. Another example is skin color. Nowadays tanned skin shows that someone has enough free time to spend the day outside sitting at a pool. This is a cue because these people cannot choose when to show their dark skin to other people. They always communicate this information about themselves. In addition this example shows that the interpretation of signs can change over time or be different among countries. In some regions in the past tanned skin showed that a person belonged to the lower classes that had to work in the fields. People with brighter skin belonged to the upper classed that does not have to work.
Signs, especially cues, can reveal internal information about the sender such as emotions, race, gender and age. Each receiver that observes a person can see his or her gender, age or race because most cues are sent constantly. Different researchers [42][168] examined this with a focus on faces and found that the social cues in the face tell a great deal about a person's internal state, or about the person herself through structure, dynamic and decoration. Mehrabian argues that 55% of the transported information in HHI is conveyed by body language when talking about feelings. Only 7% is conveyed by the word. Since body language is produced unintentionally in most cases this shows the importance of cues in HHI. Furthermore humans automatically show more facial expressions in face-to-face interaction [33](see Bruce [25] for a review of information produced by the human face)

Another important aspect is reliability. Zahavi [166] postulates that signals are always honest when the costs are high. This means that if an entity is able to put enough effort into one signal it is produced in honesty and therefore reliable. A person that has no money can not spent money buying an expensive car. If a person has an expensive car this person has to have money. Otherwise he could not have bought the car. This makes a signal honest and reliable. Maynard-Smith [104] was a critic of Zahavi's ideas. He postulated that signals do not have to be necessarily honest. Humans in particular are very good at faking signs. It is, therefore, often very difficult to figure out which signals are honest or not. Maynard-Smith postulated that there are other aspects to consider when rating the honesty of one signal. One signal is more honest if it is difficult to fake such as having a diploma. Additionally a signal is more honest if the penalty for faking it is very high. For example another way to get an expensive car is to steal it, but the penalty is very high. Renting the car for one day is less expensive and has a smaller penalty cost. This example shows that it is a very complex task to rate the reliability of one signal. Further, faking signs is not always negative and can in fact reveal other positive signals as shown in the following example taken from [39]. Many women believe they have problems with their beauty and they would give a lot to be more beautiful. The appearance of a woman is a cue and can therefore not be not produced and can be seen by everyone. If a woman thinks she does not fit the current beauty ideal she can fake it by actively changing the cue of her appearance by with cosmetics. This faking itself is a signal that tries to conceal the underlying cue. Although deceptive, this type of deception is accepted in current society. Further, everyone knows that this deception is often used, which then reveals additional signals. The quality of the cosmetics as well as the skill in recreating her appearance give additional information about the signaler.

To sum up it can be said that signs tell a receiver a lot about the sender. Yet the receiver has to decide if it trusts the signs it receives because cues may be deceptive signals or costly looking signals may be faked. If such a deception is revealed, it may reduce the reliability of the sender [104]. On the other hand, a sender that uses deceptions has to take care that the deceptions it uses are socially accepted or are of a high enough quality that the deception is not revealed.

2. THEORETICAL BACKGROUND

2.2 Perception of Robots

This thesis investigates how humans see robots and the way robot behavior influences people and their task performance. Communication between humans or between animals evolved over many years. In contrast, Human-Robot Interaction is relatively new and people are just starting to get used to it. Because of this it is important to figure out how the behavior or the appearance of a robot influences a person. This section will provide all necessary information about how robots are perceived by a human user.

How robots are perceived differs from country to country. Between Japan and the western world, in particular, there is a large gap in how well robots are accepted in society. Kaplan [81] investigated the nature of those differences. He found that in the west there is a strict boundary between the natural and the artificial. Entities can only be classified as one or the other. In Japan natural and artificial things are part of one big picture and there is no pressure to make a distinction between them. This may present a problem in the western world since social robots are artificial things that pretend to be natural. In addition Kaplan found that western people fear robots more easily then Japanese people. In Japan robots are integrated more often into society in ways that are unthinkable in the west, i.e., robots that do educational tasks (e.g. [80]). This trend can also be seen in literature and film. In the western world robots often have negative traits: they try to take over the world like in the *Terminator* movies or as in Isaac Asimov's *I,Robot*. In Japan robots are seen as more friendly and become friends with humans more often as found in the manga/anime series Astro Boy.

That robots may cause fear or discomfort was discussed by Mori [110] in 1970 with his *Uncanny Valley* hypothesis. This hypothesis says that robots become more and more familiar to humans if their human-likeness increases. But there is a gap when the robot becomes nearly human-like as it is shown in the original plot by Mori (see Figure 2.2). The idea behind this is that an entity that looks nearly human but has some strange physical characteristics or behaviors evokes discomfort because it does not fit the expectations for a human. A more artificial-like robot does not evoke such discomfort because people expect the robot to be robot-like. As an example Mori presents a dead body. It looks very human-like but it does not move at all, reminding people instead of a corpse. The dynam-

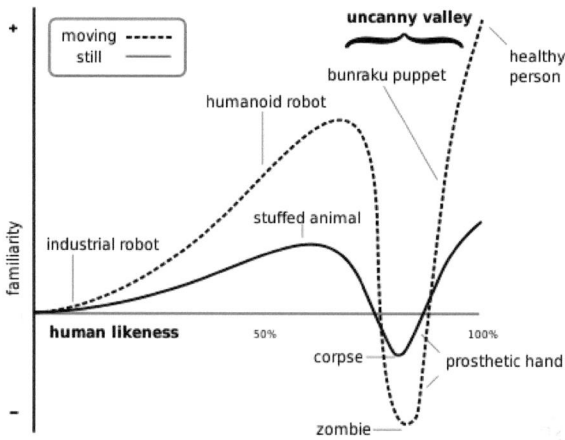

Figure 2.2: The Uncanny Valley by Mori [110] in 1970

ics even enforce this effect like a dead body that starts walking. Mori's hypothesis was much discussed in the past. For example Seyama et al. [142] postulated that the effect of the Uncanny Valley only emerges when there are abnormal features such as bizarre eyes. Minato et al. [109] added an additional dimension to the Uncanny Valley and postulate that if the appearance fits the behavior of the robot the depth of the valley can be decreased. This corresponds to the hypothesis that the form and function of a robot should fit together as it will be explained latern. It can be said that it is accepted that robots can evoke fear or discomfort if they possess distracting features. For example, a nearly human-like robot can display inappropriate behavior or abnormal features such as bizarre eyes. Minato and Seyama show that there could be a way to avoid the Uncanny Valley. The key question is how to achieve this and figure out what features are distracting and make a robot creepy.

2.2.1 Anthropomorphism

A more positive perspective to how robots are perceived is the anthropomorphism. This effect describes how humans ascribe human traits to non-human objects and beings. The most common targets for anthropo-

2. THEORETICAL BACKGROUND

morphism are animals, personal belongings and gods (see Figure 2.3). Pet owners often treat their animals as if they were humans. They compare their behavior with their own and think that they share the same needs, such as, for example wearing nice clothes. Some people love their car more then a real living person and characterize it as evil looking or cute [41]. In many cultures gods are seen as human-like entities that act and feel like people. Greek gods, for example, have strong human traits. To explain this behavior Epley et al. [49] propose that anthropomorphism is substantially influenced by two motivational factors. First people want to interact effectively with their environment(*effectance motivation*). This means that they want to understand the world they are living in. Consequently they seek explanation for circumstances they can not explain. Guthrie [63] postulates that this is the reason why gods were created that are capable of doing things humans cannot explain. The second motivational factor is that humans have the need for social contact or social connection (*sociality motivation*). People that are very lonely tend to anthropomorphism objects and animals more then socially active people. If someone has nobody to talk to their pets or loved objects fill this gap.

Figure 2.3: Example of objects that are anthropomorphised: Spoon with human-like shape beside its box (left), Stuffed animal turtle(right)

Objects do not have to be complex to be anthropomorphised. Often very simple shapes and patterns can evoke the ascription of human traits. Scholl et al. [139] wrote a review about this phenomena and showed that when people observe shapes like rectangles and circles moving across a screen, they give the shapes certain roles if they follow movement pattern that match human-like behavior pattern. For example one shape becomes the predator and the other two start hiding and protecting each other.

Guthrie formulates two major theses to explain anthropomorphism [63]. The first is the *familiarity thesis*. It states that humans use themselves as a model of the world and unknown issues because humans have good knowledge about themselves, but not about the non-human world. They seek reliable explanatory knowledge and this knowledge is the knowledge about themselves when interpreting the unknown. The second is the *comfort thesis* which argues that humans mistrust non-humans but are reassured by humans. In addition he states that anthropomorphism is a permanent effect that cannot be eliminated and thus the expectations a human has towards an object such as a robot are continuous. This means that at any time information are sent that are interpreted by a human. Duffy states that: "It is believed that language does not only function to acquire knowledge about behavioral characteristics of others, but also to find out the internal states of others (i.e. their feelings, attitudes, etc)"([44] p. 3). This is in line with the Watzlawick's first axiom for HHI: "One cannot not communicate" ([160] p.51). This means that humans always convey information about their internal state. Even when someone is sitting in a train, listening to music and not talking with anyone, the behavior alone reveals a lot about this person.

A more structured definition of anthropomorphism that extends to robots is given by Epley et al. [49]. They formulate a three-factor theory of anthropomorphism involving cognitive as well as motivational determinants. The two motivational determinants namely *effectance motivation* and *sociality motivation* were already explained above. The third factor *elicited agent knowledge* describes the effect by which humans tend to anthropomorphise more if they have more knowledge about humans, or more precise self-knowledge.

If anthropomorphism is applied to robotics Guthrie's and Epley's approaches can be summed up as that humans tend to anthropomorphise robots if they provide necessary features such as human-like behavior or appearance. Robots are strangers for humans, many do not really understand how they function or how they behave. When robots have fitting features humans anthropomorphise them by ascribing human traits and needs. With this process come expectations that the robot will exhibit human-like behavior patterns. This is why Saerbeck [135], for example,

2. THEORETICAL BACKGROUND

postulates that a robot has to move "naturally" meaning that the robot has to move in a way that is natural for the character it embodies. So the expectations can be fulfilled that are evoked by the appearance. Further, the amount of anthropomorphism depends on the personality of the human. Referring to the elicited agent knowledge each human reacts different. Some may in general tend to anthropomorphise more, others may have some background knowledge about robots which also influences the anthropomorphism. In addition the situation the interaction takes place in, or the personal situation of the person may also have an influence.

Many studies that research anthropomorphism on robots or virtual agents were carried out to figure out what features have a positive effect within an interaction or what determines how an entity is perceived. For example many robots are perceived as more friendly and human-like if they have a more human-like appearance or speech [65][84][10]. In terms of human response it has been shown that humans are more responsive when a virtual agent is more anthropomorphised, although this also evokes higher expectations of its skills [118]. Kidd et al. [83] showed that a robot is perceived as more credible and informative, as well as being more enjoyable to interact with compared to a virtual agent. In addition Lohan et al. showed that embodied robots are taught differently within a tutoring situation [93] for example by being looked at longer by the participants. Another study showed that people empathize more with a robot when it is more human-like [129]. Furthermore dynamic body movements are perceived as more appealing when they are human-like [10] and have an influence on the human gaze behavior [144]. Ter Maat et al. [99] evaluated different turn-taking strategies to find out how this influences the perception of Embodied Conversational Agents. This shows that different behavior of an entity can alter the way they are seen, and can create the illusion of various character traits. For example Agents that react very early are seen as having a strong character. Even robots that try to cheat in a game are treated more as humans when they are anthropomorphised [145], although the authors also found that some of the fraud attempts were classified as a malfunction especially when they were verbal. Riek et al. [128] collected a interesting set of comments from the interaction study they conducted. Some participants said that they forgot that they were talking with a robot after a while. Two participants in the non-mimic condition expected the robot to produce more facial expressions. Others in the mimic condition said that the head movements were too erratic or jerky indicating the expectations people have of robots. Another interesting study was carried out by Krach et al. [85]. Here the participants had to play a simple decision game against four entities with different levels of anthropomorphism, a computer, a functional robot, a human-like robot and a human. During the games participant's brain activity was recorded via fMRI. The authors found that brain activity increases, depending on the level of human-likeness, in those regions those activation should be connected to anthropomorphism.

2.2.2 The Mindlessness Approach

Beside the anthropomorphism there is another approach that tries to explain why humans react to machines in a social way. Nass et al. [115][127] postulate that humans mindlessly react socially to non-living objects. They disagree with the anthropomorphism approach, arguing that the theory cannot explain the observed effect. During their studies they found out that humans mindlessly apply social rules and expectations they learned during HHI to computers. One explanation for this might be that humans intuitively react to input they receive from the outside world independently to whether it was conveyed by a human or a computer. Furthermore this reaction is produced subconsciously and is thereby an automatic process that responds very quickly. Their main argument against anthropomorphism is that humans know that they are interacting with a machine. When people anthropomorphise, they ascribe human traits to an object. According to Nass, this is not possible because they know that the object in only a machine. Instead, people use a type of hard wired set of social rules that are triggered when certain stimuli appear. One of the main experiments in [115] examines how humans evaluate the performance of a computer. In the first condition the participants answered the questionnaire on the same computer they were judging. In another condition the questionnaire was filled out on a computer different from the one they were evaluating. In the first condition the rating of the computer was significantly better then in the second. This supports the idea that humans show politeness towards a computer when they rate it "face-to-face" as they normally do when rating a human in a face-to-face situation. An alternative way of explaining this effect is given by Dennett [40] who says that individuals frame the interaction with a computer as an interaction with the programmer and thereby address their reactions to the "human behind the machine".

2. THEORETICAL BACKGROUND

Hoffmann et al. [75] extended this research and tested the same effect on a embodied conversational agent and observed the same effects. Another study [116] associated with the mindlessness approach showed that the similarity-attraction hypothesis was also applicable to computers. The authors argue that people prefer interactions with other people with same personalities. They showed that this is also true for computers.

2.2.3 Form and Function

Independently of whether anthropomorphism or the mindlessness approach is preferred it is widely accepted that humans react socially to robots and that this effect can be strengthened when adequate features are used. In the following it will be explained how these features should be designed to improve interaction. Especially in the field of product design there are a lot of investigations about how objects have to be designed to make them usable and more attractive to consumers.

The main argument used in design is that form should suggest function [41]. For example, the spoon seen in Figure 2.3 has a human like shape and therefore feed. The existence of feed suggest that this product can stand upright. If humans see a feature that reminds them of a function, especially anthropomorphic features, they associate a certain function. Designers use this effect to guide users through an object's features to improve its usability. Another example are the signs use in robotics. If a robot has some kind of eyes a human observer will imagine that the robot is able to see and to recognize its surrounding. If the robot has facial features such as mouth and eyebrows people will expect that the robot can display for example emotion. This is the reason why, when designing robots, it is important to pay attention to the features that are presented. In other words features that evoke expectations that can not be fulfilled should simply not be there [82]. Furthermore one should make sure that signs are not ambiguous, and do not refer to different functions at the same time. If for example a robot is not moving this could mean that the robot is turned off, has a malfunction or is just not moving at all. A human standing in front of such a robot would not know if it is possible to interact with it because the signs evoke conflicting or inaccurate expectations [146]. Nowak et al. [118] formulated this fulfillment problem more dramatically. They tested virtual avatars that differed in their level of anthropomorphism but had the same competences and functions. They found that during interaction participants were less engaged when facing very human-like avatars. The authors suggest that the cause might be the high expectations they have towards the avatar which are then not fulfilled. An extensive study was conducted by Duffy where he postulated that "there should be a strong correlation between the robot's capabilities and its form and vice versa" ([43] p.185). He also collected some interesting strategies to cope with this problem. For example he proposed the use of natural motion as did Disney [148] to give characters a personality. Additionally he suggested the use of "social communication conventions in function and form" ([43] p.185) and to avoid the Uncanny Valley. To achieve this the robot should use communicational pattern adapted from humans and be as much as human-like without reaching the Uncanny Valley for example by using a more artificial design for a humanoid robot.

Function is not all that is ascribed when appraising the form of a robot. Goetz et al. [59] found out that people also associate tasks to the appearance of a robot. For example human-like robots are preferred for social tasks such as museum tour guide and machine-like robots are more easily visualized as security guards. Robots that are more serious are more suitable for serious jobs like a personal fitness trainer. This makes it clear that it is important to design the robot in a way that fits the task it will carry out. It would be inappropriate to use a 2 meter tall robot to educate children.

The appearance of a robot is very important in guiding a human user. If these features are used well they can be used to make the robot more usable and intuitive, thus avoiding frustration from unfulfilled expectations.

2.2.4 Conclusion

To sum up this section, it can be said that robots are perceived very differently depending on nationality, previous knowledge and personality. However, some effects seem to be the same for most humans. Although, sometimes people are frightened by robots, most of the time they anthropomorphise them and react to them in a social way. The key point is to find a balance between certain aspects when designing a robot [43]. The robot has to be anthropomorphic enough to be familiar to the human but

2. THEORETICAL BACKGROUND

not so much as to be uncanny. The level of anthropomorphism can be changed by providing different anthropomorphic features that can give humans guidance about the features of the robot, and reduce uncertainty towards this unknown technology. Therefore it is important to produce good quality features to gain the best effect, and avoid misinterpretations. On the other hand, these features should fit the overall function and tasks the robot should do. Features, especially anthropomorphic ones, evoke expectations that have to be fulfilled by the robot. If they are not, it may make the robot uncanny or influence interaction in a negative way. If all this is taken into account human-like features could be a great opportunity to improve HRI. Breazeal states [17] that interacting with a social robot requires no additional training because humans already know how to interact in a social way. The features suggested for improvement in this thesis are described in chapter 6.

It can be said that communication between humans and robots follows similar rules. In both cases communication is done by sending and receiving signs that influence the interaction partner. The human sees the robot and starts interpreting intentions and traits to the robotic system the same as they would during HHI. Because of this it is possible to use the same definitions of signals and cues in Human-Robot Interaction. This does not mean that in HRI the same signals and cues are used as in HHI. These issues should be investigated separately as it has already been done for many signs in the fields of emotion, behavior patter, proxemics and so on.

2.3 Arousal and Task Performance

One key factor to determine if dynamic background behavior has a positive influence on HRI is the arousal level of the human user. Arousal is connected to task performance which is best when the arousal level is at a moderate level. The goal of this thesis is to investigate how robots should behave to improve performance by influencing arousal level. If the robot stresses (though annoyance or fear) or bores the person, a decrease in task performance is to be expected.

2.3.1 Definitions

Being stressed is a natural process that everyone encounters. In particular, very stressful situations such as tests or presentations can make humans

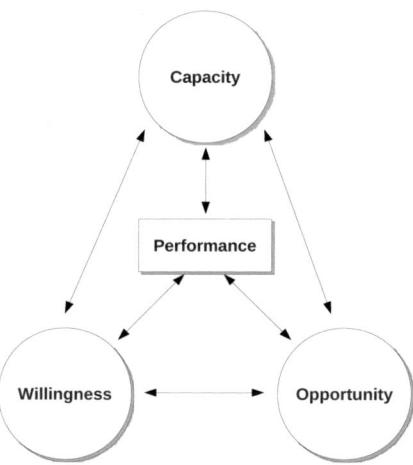

Figure 2.4: Determinations of work performance (modified according to [11])

sweat or even give them blackouts. Yet, the original reason for stress is a positive one. Stress can make people more attentive and focused when necessary to hunt or to flee. Stress can be challenging and energizing [102]. Currently stress has become a problem. For example, in jobs where stress persists over long periods of time, it can impairs attention, memory, and action. In extreme cases stress can lead to burnout, clinical anxiety and depression [161]. These findings show that stress reduction is also a health issue even in HRI. As soon as stress reaches a certain level, it decreases task performance instead of improving it.

Blumberg et al. [11] formulated the theory of work performance shown in Figure 2.4. Their model consists of three factors that influence performance. The first is the capacity (C) that encompasses all basic resources such as intelligence, learned skills and physical fitness. Willingness (W) refers to the general motivation to solve a task. The last factor is opportunity (O) which refer to the physical and social environment. They then define performance as follows:

$$\text{Performance} = f(O \times C \times W)$$

Stress influences two of these three determinants, namely willingness and capacity. If a person is stressed the willingness to complete a task decreases. Furthermore this makes it difficult to measure the influence of stress on task

2. THEORETICAL BACKGROUND

performance during user studies because humans are able to compensate stress with motivation and participants are generally motivated [102]. In addition stress has physiological influences that decrease the capacity to do certain jobs [102]. During most studies the environment is manipulated to generate the required conditions. For example in this thesis the behavior of the robot was changed to measure its influence on performance. The robot in this case is part of the environment and therefore can influence performance, for example by inducing stress. Performance itself is usually measured in speed and accuracy [102]. Only tasks of a higher quality that are solved in a short time are rated as performant.

The arousal theory introduced by E. Duffy [45] is a more detailed approach to explain performance changes. Arousal refers to a person's overall state of activity and can be defined in various ways. One example is behavioral state such as sleep or physical activity. If a person is asleep arousal is very low as compared to someone who is doing sports. A second example is emotional arousal ranging from unconsciousness to agitation. It is commonly assumed that there is a strong correlation between stress and arousal despite the current lack of evidence. If one person is stressed this increases the arousal level. But this correlation is not true in the other direction. If a person's arousal increases, it doesn't necessarily follow that the person is stressed. It can also be true that this person is emotionally aroused because she is happy or aroused because she is doing sports [12][102].

Yerkes and Dodson [165] expanded the arousal theory by adding a connection between arousal and task performance. They postulated that task performance is best when arousal is at a moderate level. The so called Yerkes-Dodson Law is shown on Figure 2.5. The relationship is designed as an inverted-U. In the middle, where the graph is at its maximum, performance is optimal. This can only be achieved if arousal is not too low and not too high. They also modeled a correlation between difficult tasks and easy tasks. If a task is easy, then it should be completed efficiently, even if the person is in a stressful situation. In the same setting difficult tasks could only be solved with a lower performance. If persons have to do a complex mathematical task, they would be more efficient if they could work in their office without any distraction from the outside (external stressors). In contrast, if someone has to do a simple task like washing dishes, this person can get bored if there are no other stressors such as music, other people and so on. It is possible that this bored person would achieve a low

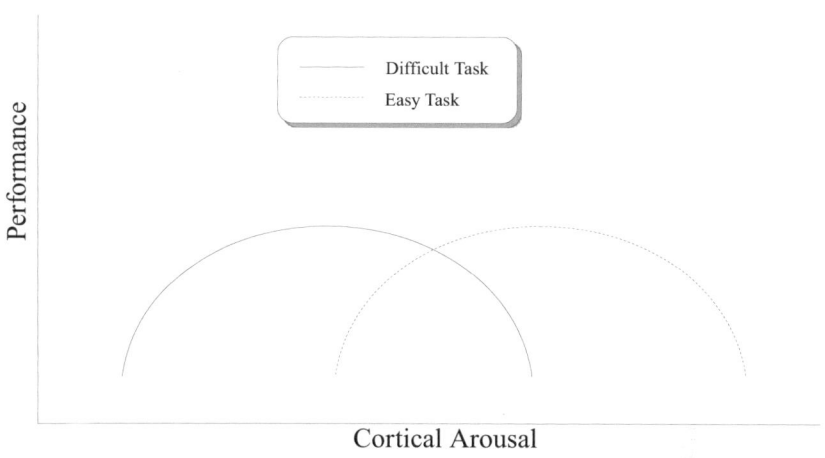

Figure 2.5: The Yerkes-Dodson Law (modified according to [102])

performance unless there were factors present to arouse or cheer her up.

The Yerkes-Dodson Law describes the correlation between arousal and performance. One weakness is that it fails to take individual differences into account [102][103]. Each person reacts differently to stress and arousal. Some react more emotionally, some are very calm or others are tired all the time and thus have a very low arousal value. Two people can reach the same performance level while having very different arousal levels. This makes it difficult to use this measure to compare performance by measuring arousal. This problem is addressed in section 2.4 and 8.2.2.2. In addition Mattews et al. [103] postulate that stress can be split into three dimensions: (a) Task Engagement refers to the motivation to solve a certain task, (b) Distress describes the feeling of loss of control and (c) Worry is a conjunction of different self-references such as self-consciousness or low self-esteem. These three dimensions define personal stress for one special task. In other words, each person perceives stress in a different way.
Because of this external stressors can have a different effect on the arousal of each person. Stressors are defined as external influences that alter the arousal level [102]. There are many different stressors ranging from simple noises caused by cars driving by, to talking people sitting next to someone trying to concentrate. For example, people such as firefighters or astronauts

2. THEORETICAL BACKGROUND

who work in extreme conditions must learn to cope with disturbances that affect their arousal.

Everything that influences arousal extrinsically is defined as an external stressor. As such, robots also fit in this category. This means that there is a correlation between the robot and task performance. Depending on the robot the effect on arousal differs according to its appearance and behavior. Here the focus lies on how dynamic background cues can influence the arousal level of the user. Optimally the robots should either arouse or bore the user too much. Interaction with a robot is normally arousing because most people are not familiar with robots and cannot predict their behavior. This project investigated whether certain behaviors could reduce arousal.

2.3.2 Humans and Robots in Cooperative tasks

Cooperative tasks with robots are a special kind of interaction. A typical interaction does not require a given outcome. Cooperation differs in that all participants have a goal. These goals do not have to be the same for both. Further, a cooperation can take on different structures. First, there are tasks that can only be completed if both parties work together as when lifting heavy objects or playing a game. Second, there are tasks that can be completed by one entity alone where the second entity assists or is in a passive role. This might be for example a surgery nurse that assists a surgeon or a teacher that teaches a pupil. These cooperative tasks are useful for testing task performance, as well as social effects in a cooperative setting in HRI.

Several of the effects described above can be applied to face-to-face cooperation. Most of the time the human and the robot are facing each other. In this situation, a robot can be seen as a stressor, because it cannot be ignored. It also facilitates mutual communication. Cues in particular have a high value even if there is no speech. Because of this, anthropomorphism or other social effects can be evoked, as described in section 2.2. We posit that social aspects have high relevance in cooperative tasks and can improve these interactions.

Referring to Blumberg [11] the robot is not the only stressor in a cooperative situation. Other factors can influence performance as well. At first sight the task itself determines performance and produces a certain arousal. Another factor is the situation the person is in. Are there any disturbing stressors such as heat, noise or bad lighting conditions? Does the environment support or interfere with the task? That said, the primary goal here is to investigate the influence of robots on people, with a focus on stress and arousal.

In the following part, studies will be reviewed that focus mainly on Human-Robot Cooperation and Stress reduction. Robots are able to improve cooperation in many different ways. Two studies [38][1] that focus mainly on usability found out that task performance can be improved and that stress can be minimized if the robot is easy to use. Robots that are difficult to steer or use unusual control schemata can be unusable. Although these studies were not conducted with social robots, they support the findings of Luczak et al. [97]. The authors argue that machines that have malfunctions or produce unreasonably large delays during interaction are too stressful to users. So much so, that it is not uncommon for users to shout at or hit the machines in an effort to reduce their stress.

Another strategy to improve interaction is for the robot to help the user directly to complete the task. Some robots use more obvious strategies such as giving hints. Chan et al. [32] investigated the effects of stress depending on whether the person was doing the task alone or in cooperation with a humanoid robot that gave constructive hints. They found that for tasks with high task-induced stress a social robot could reduce stress by providing help.

The other strategy is to help the user by giving unobvious hints. For example Mutlu et al. [114][113][76] did a lot of research on robot gazing. Nonverbal leaking, as defined by Eckman [48], was used to give supposedly unintended hints to help the human user. During the studies the participants were asked to play a guessing game in which the robot sometime "unintendedly" looked at the object that should be guessed. This helped to finish the game more efficiently than without nonverbal leaking. Breazeal [20] used social cues within a teaching task where the robot gazed (or not) at the learned object. Participant had fewer problems teaching the robot the name and the location of the buttons if the robot focused its point of attention or pro-actively displayed confusion. Another study [57] showed that robot motion that could be anticipated by the user improved the interaction. Users were able to understand the intention of the movement earlier than without anticipation, providing them with more time to react. Furthermore it has been shown [60] that empathy towards a robot increases the perceived subjective task-performance.

2. THEORETICAL BACKGROUND

Often it is not important how the robot behaves. The appearance or presence of a robot can be as important. Rither et al. [130] applied the social facilitation study [167] to robots and found that the mere presence of a social robot can increase task performance for simple tasks but impair performance of difficult tasks. The seal robot Paro [136][153] is capable of reducing stress of elderly people just by being present. In addition it has been shown that humans rely more on robots that appear more human-like [74], and that they prefer to cooperate with more serious behaving robots [84]. Eyssel et al. [51] proposed that it was important to design robots in a way that it reduces stress by giving humans the feeling that they had control within the interaction.

Much work has already been done in the field of task performance and on the influence of arousal/stress level in humans. What is missing is structured research on simple social cues that have no functional meaning within the interaction. Often social cues like breathing and blinking are used in HRI to make a robot more social. However, there is no research on what influence these single cues may have if they are considered separately and independently from other social cues that actively improve task performance such as nonverbal leaking.

2.4 Measurement of Stress

Often the term "stress reduction" is used as a measure for the quality of HRI. Unfortunately there is no way to measure stress directly. Biophysiological measures provide the opportunity to directly asses arousal levels in connection with the stress. In general biophysiological measures make it possible to assess internal states of the human body such as stress, anxiety and affective state. The advantage of this method is that it measures direct body responses of the participant. Therefore more objective data can be collected compared to, for example, questionnaires. To measure arousal *Electrical Skin Resistance* (ESR) and *Electrocardiography* (ECG) are often used because these two values are connected to the internal nervous system and are not influenced by the thoughts of the participant. Another very popular biophysiological measure is the *Electroencephalogram* (EEG) which measures brain activity. The disadvantage of using ESR and ECG is that it is only possible to measure arousal level which has to be interpreted afterwards. An increased arousal level can indicate stress as well as happiness[12][102]. Kulić et al. [87] call this additional dimension valence, for which the value is gathered by having participants complete a self-assessment questionnaire, or by using Electromyography (EMG) of the corrugator muscle. This distinction in two dimensions is inspired by the valence/arousal model for emotions described by Russell [134]. To interpret ESR and ECG it is essential to use a questionnaire in addition to getting a sense of participant's subjective impression of the situation. Bethel [9] provides a very compact overview of this topic.

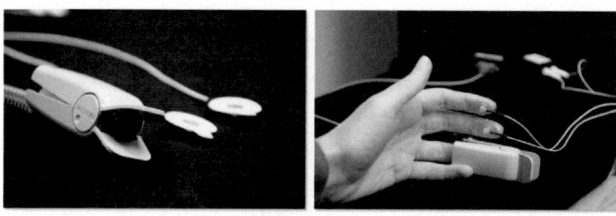

Figure 2.6: EKG and ESR Sensors

Sensors for ESR and ECG are shown in Figure 2.6. The sensor that is attached at the ring finger is the ECG, the other two are used to measure ESR. These sensors are connected to a recording device which sends all sensor data to a computer where the information are stored. Technical, ESR is nothing more than the conductance between the two sensors. Depending on sweat production the conduction changes. Sweat production is directly connected to arousal [9][126]. Two components are involved in the stress reaction of the human body [30] the hypothalamic-pituitary-adrenal and the sympathetic nervous system. Most organs are connected to the sympathetic nervous system as well as to the parasympathetic nervous system, with the exception of the skin [12]. The skin is only connected to the sympathetic nervous system which directly regulates arousal, and therefore boosts the performance of the connected organs. The parasympathetic nervous system is the antagonist to the first whose function is to calm down the organs. This structure is an advantage when using ESR [141]. Because the skin is not calmed down by the parasympathetic nervous system (like the heart), it can provide a direct mapping of arousal. In addition skin conduction responds very well to emotions [12] independently of whether they are positive or negative. For example some people get sweaty hands in inappropriate situations such as when they fall in love. The ESR response is a relatively slow reaction, a respond can be measured on average 1.6 - 6.5

2. THEORETICAL BACKGROUND

seconds after the stimulus, decreasing slowly until it reaches the relaxed level again [141]. When several stimuli are presented one after another, the reactions interfere with each other creating a typical arousal plot as shown in Figure 8.5. Here the value that is recorded by the ESR sensors is plotted over time. This plot illustrates nicely when each stimuli becomes visible by a peak in the plot. The reaction of one participant during the interaction study presented in section 8.2 is shown in this figure. The three segments correspond to three tasks the participant had to complete.

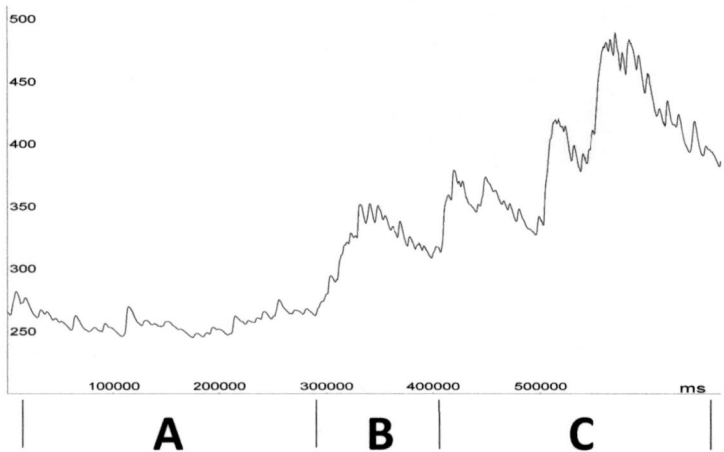

Figure 2.7: Sample ESR plot. Low values represent a low arousal level. Segments A-C stand for different tasks that had to be completed one after another

The ECG measure behaves similarly to the ESR. The heart rate increases when the participant is aroused and decreases when she calms down. This makes ECG a good method to validate ESR measurements. Furthermore both methods can be measured in parallel quite easily. They both only need 1 or 2 fingers, and do not disturb the participant's movements very much, as they only need to hold the hand still. One problem with using ECG is that the heart is not only connected to the sympathetic nervous system. It is influenced by many other mechanisms within the human body. For example, heart rate decreases when a person holds their breath [37]. In this situation arousal increases because of the lack of oxygen which can be measured by a increased ESR value although the ECG reports a lower heart rate. This shows that ECG is less reliable but still a good support for ESR.

Biophysiological measures have large advantages. They are objective, non-invasive methods, meaning that the measure is not influenced by culture or education and can not be willingly manipulated by the participant. Furthermore it is not necessary to injure the participant to attach the sensors. There are also some disadvantages that need to be addressed. As mentioned before there are individual differences between subjects. Two ESR values do not say anything about the arousal of two different people [73][102]. A few methods have been developed to compare different participants. The most common is to record a baseline when the person is relaxed and compare the differences between the relaxed and aroused values [73][9][143]. In addition ESR and ECG can be influenced by caffeine, nicotine and drugs making it necessary to instruct participants not to consume these substances directly before the experiment or at the very least ask them if they were consumed [9]. In addition there is the possibility that ESR does not show a normal response for some people. This is not due to an illness. Instead some people simply do not have normal skin conductivity.

To make biophysiological measurements more interpretable it is recommended to use a questionnaire. The sole use of ESR and ECG make it difficult to know whether the arousal was caused by stress, happiness, sporting activity or something else. Bethel et all. [9] recommend the use of self-assessment tests like the PANAS scale to test emotional arousal or valence for each participant during an experiment. The PANAS-X scale [159][158] consists of 60 items that are assigned to 13 categories such as positive, negative, fear, joviality, etc. This test can be used to determine what emotions the participant had, with the ESR and ECG indicating the strength of the emotion. In addition other questionnaires can be used to investigate how the participants rate the situation. One example is the Godspeed anthropomorphism scale [6]. section 8.2 deals with the biophysiological measures, and addresses how problems were resolved

Biophysiological measures have been used successfully in different robotic studies. For example Rani et al. [125] and Kulić et al. [87] investigated how computer or robotic arms affect anxiety in humans. They did this by using a robotic arm that performed some movements near the participant. In one condition they used a standard potential field planner, in the other a save

2. THEORETICAL BACKGROUND

planner that would behave more socially by e.g.: keeping more distance from the human. In this study they showed that humans felt less anxious during the save condition. In other studies [138] [90] it has been shown that robots are capable of detecting human affective states and to adapt their behavior accordingly. Stress reduction or stress detection is one of the most studied fields when using ESR or ECG in Robotics. Chan et al. [32] (s.a.) for example, also used ECG to measure stress levels during a task to be completed. If the participants were stressed by the task the robot provided more hints to complete the task more easily, resulting in reduced stress levels. Other studies [126] [73][143] investigated different strategies for online stress detection and compensating for individual differences to influence the robot's behavior. Others [77][141] focused more on inventing wearable devices to measure biophysiological data without disturbing the human. For example [77] developed a suit for the upper body that measures heart rate, respiration, perspiration, pulse wave, blood pressure and arm motion by allowing unimpeded motion.

Chapter 3
Definitions and Hypotheses

In the last chapter it was shown how humans communicate, and the ways robots can influence the human within HRI. The focus was on how robots can influence a person through their appearance and behavior. There are many different signs that can influence how humans perceive robots. It is sometimes difficult to keep track of all the relevant ones. In section 3.1 a typology is introduced that was created by some colleagues and myself [72]. This makes it possible to organize signs that are produced by a robot so that it is easier to cluster different types of signs and to structure research more efficiently. This facilitates the process of identifying which signs influence the user and which may influence each other. For example dynamic signs have no effect when a person looks at a picture of the robot, and olfactory signs have no influence on video.

Dynamic Background Cues (DBC) are defined in section 3.2. These cues can be sorted into this typology and are therefore a set of signs that can be used to influence the perception of the user. The goal of this thesis is to create a set of DBCs that improves HRI . They are designed in such a way to help make the robot more lively and human-like. The study will investigate if this influences familiarity as well as performance. The detailed hypotheses and a discussion as to how HRI could be improved by using DBCs are given in section 3.3.

3.1 Typology of Signals and Cues

The main goal in creating a typology of signals and cues for robots is to better structure the broad variety of signs that can occur during HRI. Signs can be of different types and can be sent by many sources. Often all signs are seen as having the same type. We think that it is useful to distin-

3. DEFINITIONS AND HYPOTHESES

guish between different types of signs on the one hand to better discuss and compare signs with other researchers, and on the other hand to better categorize them for source, types, etc. Different types of signs also affect a human observer in different ways. Visual signs, for example can be seen over large distances whereas olfactory signs are mostly only perceivable when standing near the robot. Another useful feature is the possibility of clustering different signs in groups of signals of the same type. In addition this typology helps to identify signs that the designer was unaware of.

The typology that is presented here was originally designed by some colleagues and myself [72], and was inspired by the theory of Maynard-Smith [104]. We modified it for this theory to better cover some aspects that were missing in the original version. Maynard-Smith already distinguished between signals and cues in animals. As a reminder, signals are intentionally given signs, whereas cues are unintentionally given signs in communication. Signs itself are the most general unit in communication and describe each information conveyed by a sender. More details of signals and cues are given in section 2.1. This typology adapts the theory of Maynard-Smith to robots and adds new dimensions that allow us to categorize each sign in 4 dimensions. These dimensions are illustrated in Figure 3.1, and are explained in detail in the following chapters. Since this typology consists of dimensions that are strongly connected to technical devices (e.g.: artificial vs. human-like) it is no longer applicable to Human-Human interaction. This definition focuses on communication between humans and robots.

To use this typology efficiently, another aspect has to be mentioned. Signs are not the same as an action that is carried out by a robot. Performing a pointing gesture is not a single sign but rather consists of many different signs that in combination form this movement. Most of the commonly used behavior in robots can be disassembled into many single signs. This step is often the most difficult one when interpreting and classifying robot movements. The next two examples will clarify this topic a bit more. The appearance of a robot is very good at showing that signals are more differentiated than they appear at first sight. Their appearance cannot be seen as one single sign. It must rather be divided into many small parts that comprise the overall appearance. The eyes can be one cue as well as the mouth or the fingers. Depending on how detailed the signs are, a single sign can be divided into even more signs. For example the eyes, as one cue, can again be divided into more cues such as eyelids and eyeballs.

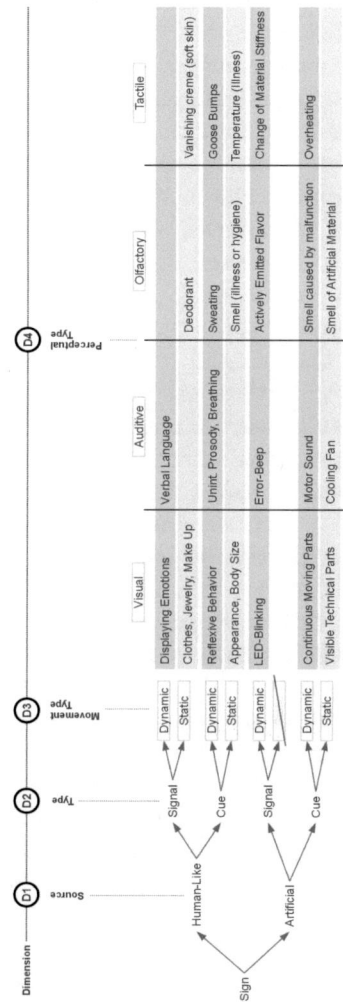

Figure 3.1: Typology of human-like and artificial signals and cues for robotics. Examples for all dimensions are given in the lower right section

3. DEFINITIONS AND HYPOTHESES

Even the motion of each single muscle can be seen as one sign. This degree of fragmentation can not be handled in most cases when analyzing the affects on the recipient because these influences are too weak to be measured. In recent literature a less detailed fragmentation is common, e.g. whole parts of face such as mouth or eyes, whole movements such as performing a step forward, or gazing movement are considered. In this thesis we used a similar fragmentation, but in some cases divided a sign further to a degree where it can be clearly classified in classes described in this section. For example, The mouth is divided into its dynamic and static signs when needed. A sign can only be classified to one certain class if all containing signs are of the same class. In other cases this sign belongs to multiple classes to a certain ratio, i.e. the mouth in whole belongs to the dynamic as well as to the static signs.

Furthermore, cues like the material of the robot can overlap with other cues. If different parts (and therefore other cues) are made of the same material the material can be seen as one cue that is separate from the cue that it overlaps. All these cues form the overall appearance. This again shows that it is important to design signs in such a way that they fit each other (See section 2.2.3). Two signs may have positive effects when observed separately but may cause discomfort when they are used in parallel. If they do not fit together they can provoke expectation mismatches as when using happy eyes with an angry mouth.

The second example is the previously mentioned pointing gesture. Here, dynamic and static signs are involved. The most obvious sign is the pointing itself. Aside from this signal, other signs can be observed. The way that the pointing is carried out is a dynamic cue that can convey information about the health status of the sender by being strong and healthy, or weak and ill. When dealing with robots, there might also be noises caused by the motors used in the joints. Further, the appearance of the arm that performs the gesture is also involved.

3.1.1 The human designer

The human as a designer of robots is not really a dimension in the topology but remains an important factor for robotic signs. Nearly all robots are designed and build by designers, mechanics and roboticists. These people generate all the signs a robots conveys. Some of them are explicitly designed by them, others occur without being designed. The jerky movement of a robot's arm is often unwanted and occurs by accident. Regardless, this movement is a dynamic cue and will be interpreted by another human.

When seeing the robot as a product of roboticists it is not obvious why a robot should produce intended or unintended signs because all signs produced by a robot are intended from the viewpoint of the roboticist. In this case, all signs from a robot would be signals. To deal with this, the typology covers the so called robot's view. Here the robot is seen as an autonomous entity that can make decisions on its own and therefore can act with and without intention. As an aid for classification, the robot can be seen as a human and the signs are classified as if they were produced by a human. This makes it easier to interpret robot signs correctly. Another reason why robots can be seen as acting entities is that robots are becoming more and more autonomous. Often the behavior of robots is produced by using probability functions or neural networks. All of this makes a robot independent from the roboticist. They can react to external stimuli making their behavior not longer completely predictable.

As mentioned above, there are two types of signs when looking at robots from the roboticist's point of view. On the one hand, roboticists explicitly design robot behavior and appearance to achieve a certain goal. This can be a special function the robot should carry out, or a look the robot should have. Beside these explicitly designed signs, there are signs that are not explicitly designed by the roboticists. According to 2.2 robots send information all the time, and it is often the case that aspects of the robot are interpreted even though these signs are oversight or generate unwanted information because of bad designed by the roboticists. Explicitly designed signs are all those that are defined as such by roboticists, and that are programmed, build or animated to reach a certain goal. Depending on the quality of a sign, these can have a positive or negative effect on the interaction. For example the jerky movements that occur from a defective control in a joint can make the robot uncanny because the movements remind people of diseases like Parkinson. In addition, combining two high quality signs can have negative effects when they do not fit together. Non-explicitly designed signs are all those signs that occur without being intended. They can be, for example visible cables, static cues that were forgotten during the design process like missing ears or joint covers. There is a large variety of unintentional signs, most of which should be actively avoided by designers at the risk of creating an uncomfortable user experience. That is why We advise to minimize these signs as much as possible to improve interaction by minimizing expectation mismatches.

3. DEFINITIONS AND HYPOTHESES

Another necessary topic is deception. When designing a social robot, roboticists make use of human-like signs which do not naturally belong to robots, making them, by definition, deceptive. As already discussed in 2.1 deceptions that are revealed can decrease the credibility of the sender. On the other hand in following with Nass [115], humans react socially to robots even when they know that they are interacting with a robot. In addition, the seen in prior studies, humans accept these deceptions to a certain degree, the same as in HHI (see the make-up example in section 2.1).

3.1.2 Dimension 1: Source

The first dimension of the typology is the source of a signal. The source can be either human-like or artificial. Signs are defined as human-like when they adopt human behavior or appearance. Artificial signs are those whose source is of a technical nature. In principal, all conventional symbols and artificial interfaces used in products which automatically represent a state are by definition artificial signals or cues intended to alter the action of a user. For example in the field of car design sound designers at Porsche explicitly try to improve the perceived quality of the car's functions by modulating the sounds of motors, clicks, and sound intensity [61]. For social robots, often human-like, as well as artificial signs are used. Depending on the design of the robot, there may be more of the first or the last. For example the Geminoid has nearly no artificial signs. On the other hand, the iCub has many. It is often the case that artificial signs are used explicitly to avoid the uncanny valley. As such it is not necessary to design a social robot to be as human-like as possible. Duffy states "The human form and function is not the ultimate design reference for a machine, because it is a machine and not human" ([43] p.185). Because of this it is reasonable to integrate the source into the typology. As an additional source, it is also reasonable to integrate animal-like signs as used for PARO, although this discussion lies beyond the scope of this thesis.

Using artificial signs in social robots allows us to convey information that is not possible by using human-like signs. Currently most humans are familiar with the signs that are used by computers. Blinking LEDs are used to represent for example hard disk activity (Figure 3.2(left)). Using these signs allow us to show similar processes in HRI such as memory overload, overheating, direct display or internal states (e.g.: thinking, ...).

Figure 3.2: Examples of artificial signals: Blinking status LED of a computer (left); BERT2 with screens instead of eyes and mouth (middle); Mobile robots BIRON (right)

The hybrid expressive face BERT2 [7] for example has three displays that replace eyes and mouth. With these it is possible to display human-like signs by displaying eyes but also to use artificial signs such as iconic symbols like hearts to show that the robot is in love (Figure 3.2(middle)). Further, Peters et al. [120] found that users prefer to have visual indicators on the mobile robot named BIRON to signal them in which direction it was going to drive next (Figure 3.2(right)).

3.1.3 Dimension 2: Type

The second dimension distinguishes between signals and cues. This is the classical dimension that is originally introduced by Maynard-Smith in [104]. The meaning of signals and cues has already been explained in section 2.1. In this typology they are adapted to Robots in HRI. For robots the definition is the same as for humans. The robot's point of view is as described above. Signs that are intentionally produced by the robot are defined as signals that are produced to influence the human observer actively. For example, pointing gestures or speech fall in this category. Cues, in robotics, as for humans, are all those signs that are produced unintentionally. Since robots do not produce unintentional signals, because every behavior has to be produced actively, human standards are assumed (robot's view). Therefore idle movements like breathing or blinking as well as the appearance of the robot become cues.

3.1.4 Dimension 3: Movement Type

The third dimension describes whether a sign is dynamic or static. This means that there was a separation between signs based on motion or exis-

3. DEFINITIONS AND HYPOTHESES

tence. Dynamic signs are defined as signs that convey information through motion. Returning to the pointing example, a movement can be classified as strong or weak. Depending on how the arm moves, different information about the internal state of the robot is produced. These are dynamic signs. In contrast static signs are those that convey information by mere existence. All signs that form the appearance of a robot are in most cases static. They do not change, and affect the observer just by being there. In some cases, it is difficult to unambiguously classify signs as static or dynamic. For example, some diseases change the color of the skin. The color changes over time, although during an interaction the change is not noticeable. This makes the sign static because skin color is perceived as constant during the interaction.

Donath [42] gives a similar definition for human signs. She used three types: structural, dynamic and decorative. Dynamic signs in her definition are the same as in this typology. Structural signs are those that are "genetically determined features such as bone structure and skin color to assess someone's personality, morality, intelligence, etc." ([42] p.6). In other words everything that is defined by the human itself. Decorative signs are signs that decorate the structural signs and "reflect one's choices and circumstances" ([42] p.6). These can be, for example, glasses, clothing, makeup or hairstyle. For robots it is impossible to distinguish between structural and decorative signs because the whole robot can be conceived of as a type of decoration that is produced by the roboticists. As such, structural and decorative signs can both be considered static signs. For similar reasons there are no artificial static signals in this typology (Figure 3.1). Artificial static signs are always cues because they are all designed by the roboticist. The robot cannot intentionally change its appearance because it has no fixed appearance as humans do. Following from Donath, robots can be said to have no structural signs only decorative ones. A change in a robot's appearance becomes its new real appearance and is therefore again a cue.

Emotion displays can be useful when explaining the difference between dynamic and static signs. First the emotion itself is static since it does not change at the moment of observation. Often emotions are not persistent; emotions can change during an interaction. When this happens, the change of emotions is a dynamic sign. The rate of emotional changes, the speed, or the emotions involved are all sings separate from the static emotion first perceived. This shows that static signs do not have to be fixed during the whole interaction. They can change many times, although the crucial point is that this sign is interpretable by it's existence independently of how long it is shown. A good illustration is that emotions can be interpreted in photos, but it is impossible to interpret the transitions between then. This is the same for all dynamic/static signs. If the effect takes place in a photo of the entity, then it is a static sign.

3.1.5 Dimension 4: Perceptual Type

The last dimension does not refer to what signs are sent by the robot. It focuses instead on the senses that are necessary to perceive the sign. A human has 5 senses to gather all information about its environment. For HRI four of them were selected to classify signals and cues. Those are visual, auditive, olfactory and tactile. Taste is left out, because this sense is rarely used in HRI. The most common senses are visual and auditive, but the olfactory and tactile senses are also relevant despite being neglected in current research. The appearance of a robots is a classic example of a visual sign. All signs that can be observed by using the eyes are visual signs. Auditive signs are mostly used when speech is a part of the interaction. Unwanted noises like cracking joints also fall in this category. Tactile signs are those that can be perceived by touching the robot. A good example of this is the PARO Seal robot [136]. The hide of this robot was designed to be very soft and to emit warmth to make it more lively. The olfactory sense is not used in robots as far as We know, whereas already devices that are capable to synthesize or detect scents exist. Such devices as well as static scents could be used to achieve a certain effect in HRI. On the other hand it could be disturbing to use a material that smells ugly in interaction. Further, uncommon smell such as smell of burning could be a sign of malfunctions.

3.2 Definition of Dynamic Background Cues

The research described in this thesis mainly focuses on Dynamic Background Cues (DBC). This chapter will describe how they are defined and what advantages they have. DBCs are a special type of sign in that they can be classified as dynamic cues following the typology, and that they occur in the fourth dimension for all senses. In other words, DBCs are signs that are produced unintentionally and are defined by their movements. Hu-

3. DEFINITIONS AND HYPOTHESES

mans produce many of these signs for example when breathing or blinking. Many of these are reflexes integral for the human body. Although not necessary to robots, they help to make robots more human-like and lively. In this thesis DBCs will only focus on human-like features although artificial signs can be used to make the robot more lively. As such, all unintentionally movements produced by a human are potential DBCs because they are natural to humans. If they are missing in an anthropomorphised robot this could lead to expectation mismatches possibly even to the Uncanny Valley.

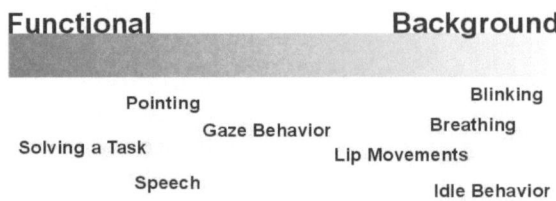

Figure 3.3: Importance of functional and background signs for different movements

The main aspect of DBCs is described by the keyword background. Beside the typology signs are divided into functional and background signs. Functional signs are defined as signs that directly help to complete a task. On the other hand, background signs are defined as the opposite. These signs do not have relevance to a specific task, and provide no additional task relevant information. Because of evaluating how the robot is perceived only those information are relevant for the classification that help the receiver in this particular task. Of cause for humans it is essential that the observed person breathes because without this behavior the person would not survive and therefor not be able to solve the task. For the classification breathing is still background because it does not provide additional information that help to complete the task such as doing a puzzle, even though being a requirement that both persons are able to solve the task. Background tasks are therefore often reflexes and idle movements that are driven by physical needs (breathing, blinking,...) or internal states (mental state, state of health,...). For example, a direct manipulation necessary to putting dishes into the dishwasher is a functional sign. In contrast, unintentionally playing around with a pen as an idle movement is a background sign even though the same body parts are used. Another example is breathing. Since it has no relevance in a cooperative manipulation task, it is considered a background cue. On the other hand, breathing rate may become relevant if a person wants to figure out if another person is lying. In this case the same cue becomes a functional cue. This shows that the definition strongly depends on context of the task, and explains why the distinction between functional and background signs is not integrated in the typology. The typology only describes the signs of robots without considering the task.

In addition to the signs described in the typology, functional and background signs also occur in parallel when considering a behavior. Referring again to the pointing gesture, the pointing signal is the foreground part of this behavior. It directly refers to the function of the gesture which is to direct attention to a given place or object. Other signals emitted in parallel (which are mainly cues) only have background functions in this task. The appearance of an arm or the way the pointing is carried out have not relevance, and provide no additional information. The pointing is the most important sign for this behavior because it is its main aspect. That said, background features have an effects as well. To give another example, using make-up to change the appearance would be a background signal for the pointing task because it has no task relevance. Further, when using verbal language in HRI, speech itself is a functional sign that transports information to the recipient, but the movement of the lips would be a background sign. Figure 3.3 shows some frequently used behaviors in HRI. These are placed depending on the ratio of functional and background signs for a common cooperative interaction. Behavior plotted on the right has an important background component. For behaviors located on the left the functional component is of higher importance. Thereby only the importance is crucial, not the amount of background or foreground signs

Often robots use a large quantity of functional signs and have only a small amount of background behavior. The goal is to make a robot more social by raising the amount of background signs. The distinction between functional and background is motivated by the attempt to find a way to make a robot more social without influencing the functions of a robot too much. Often robots already have functions and behaviors, such as grasping objects , that have already been implemented. The aim was to make the robot more lively without having to change the implemented software and behavior. The theoretical definition of DBCs was born out of this goal, making it possible to decouple the semantic and an implementatory levels. For the implementation this allows the creation of a type of social layer

3. DEFINITIONS AND HYPOTHESES

that can be put between the functional part of the robotic system and the robot as hardware which realizes the integration of DBCs. For example, the functional commands can be made more social or some additional movements can be added. In section 7 it will be explained how this layer is implemented.

This definition makes it possible to investigate the effect of DBCs on the human user, especially with respect to arousal and task performance, as they do not have any functional influence on the performance by way of hints, etc. By using only DBCs, it is possible to examine the social aspects in HRI. The DBCs that are used enrich the existing behavior of the robot by adding a smooth movement behavior onto an existing functional arm movement or adding a simulation of breathing onto a stiff robot posture. Many of these behaviors have already been used in movies or in other robots, although they have never been formally investigated. Often the influence of single cues in the humans are not known but simply adapted from humans and used on robots.

This thesis often uses "social behavior", "social movements" or similar terms. These terms refer mostly to DBCs or other social dynamic features. Unfortunately the term social is not clearly defined, which is why a typology is used to have a more structured definition of these signs. Further, the task that is assumed here is a general interaction between a robot and a human that mainly focuses on completing a cooperative task (e.g.: a tutoring task). Therefore, the DBCs that are explained in chapter 6 were all chosen because they are background signs in such a situation as defined in this section. The assumption to classify these DBCs as background within the given task can be made because the conducted studies, observation of users interacting with the robot and recent literature support this. Further, the presented investigations proof that this assumption reveals consistent affects and therefore can be made.

3.3 Hypotheses

Using well-placed and well-designed signs makes it possible to reduce the amount of expectation mismatches hopefully leading to better robot interaction. The goal of this thesis is to investigate if it is possible to improve interaction with robots by using only very subtle signs (the Dynamic Background Cues) that have no influences on potentially existing functional behavior. The experimental measures used are correct interpretation of social features (familiarity) and change in the arousal level of the user (task performance).

Out of this the following hypotheses are formed.

Hypothesis 1. *Increasing human-likeness by using DBCs increases familiarity and perceived liveliness of the robot within a cooperative task.*

This hypothesis underlies a proof of concept hypothesis formulated to investigate whether DBCs in general take effect in HRI. Furthermore, this investigation was used to get feedback and experience on how DBCs take effect and how they can be improved. Heerink et al. [68] research is used to postulate that a more socially expressive robot increases comfort and therefore familiarity. Further, of central interest is whether the robot is perceived as more human-like and is therefore anthropomorphised. The main study addressing hypothesis 1 is described in chapter 5

Hypothesis 2. *The use of parameterized DBCs makes it possible to create certain robotic character traits that are perceivable by human users both when using single and multiple cues in combination.*

The investigation of this hypothesis will show if it is possible to form certain character traits such as reserved, happy or nervous by using simple dynamic cues in humanoid robots. Depending on the task different traits are expected or accepted by the human user. The question is whether single cues are able to form these traits depending on their parametrization. Furthermore, this investigations addressed whether these outcomes could be combined without disturbing the overall perception of the robot. Having these cues helps to make the internal state of the robot perceivable as a social display. During several online studies, data was collected to see what effect single DBCs had. These outcomes were used to design a robotic character that could carry out the necessary functions during the target studies. The online studies are described in chapter 6 and 8.1.

Hypothesis 3. *Minimizing the amount of expectation mismatches by using DBCs decreases the arousal level of a human user within a cooperative task.*

The idea behind this hypothesis is that a robot that minimizes the amount of expectation mismatches is more familiar and therefore stresses

3. DEFINITIONS AND HYPOTHESES

the user less than if there are more mismatches. This would mean that the arousal level should decrease when the robot's behavior is more social. As arousal is connected to task performance, this could mean that a social robot facilitates completing a task in cooperation with a human. To measure arousal biophysiological measures were used to gather objective data from the participant. The study that investigates the third hypothesis is presented in section 8.2.

In summary, if this hypotheses can be supported, DBCs should be able to increase familiarity with a robot, and manage to alter the arousal level of a user so as to improve task performance without influencing the underlying functional system of the robot. Furthermore the robots would appear more human-like by only using very subtle DBCs. To achieve this, all DBCs should be aligned to each other so that there is no disturbing inferences between them.

In contrast to other research in this field, in this thesis the functional and the background signs are strictly separated. Other studies have often mixed them up and evaluated both in parallel on familiarity or task performance. Limiting the focus to background signs provides a deeper view into the effects of pure social signs, especially when evaluating them alone and in combination.

Chapter 4
Review of relevant Social Robots

In the past years a variety of social robots were created by different roboticists, despite the difficulty. The last chapters showed that there are a many problems that can occur when designing a social robot, and that it is a challenge to build one in a way that is accepted and loved by people. Some interesting current strategies have evolved to cope with the previously discussed problems and to make the robot more familiar and friendly. Fong et al. provides a good review of social robots [53].
This chapter introduces design criteria for social robots. A review of social robots will follow including an analysis how the robots fit to these criteria. Further, it will be discussed what strategies other roboticists used to make their robots more familiar. The reviews will be followed by the implications for this thesis, particularly in terms of system design and creating adequate DBCs. Analyzing foreign robotic systems with a focus an the four criteria is very useful when designing one's own system. For example the first set of DBCs presented in chapter 6 was inspired by different systems that already used similar cues. The same is true for the system implementation presented in chapter 7 which combined some classical approaches and extended them to the final system described in this thesis.

The related work presented here only describes considerable social robots, and the ways they influence the design of DBCs and the robotic system. Related work focusing on the theoretical basis of DBCs can be found in chapter 2 and 3.

4.1 Criteria for Social Robots

The four criteria for social robots that are introduced here are key aspects relevant to a robotic system and should be noted when creating the design.

4. REVIEW OF RELEVANT SOCIAL ROBOTS

When setting these up it is advisable to be familiar with the definition of social robots. Breazeal formulated a set of questions [17] that can be used to get an impression what the important topics are when designing a social robot. The set contained questions such as "what expectations should a robot fulfill" or "for which cultural region should the robot be designed". These questions originally directed researchers to begin a discussion of what research was relevant for HRI.

Breazeal and Duffy gave two definitions that will be presented here. Breazeal's definition of social robots consists of four categories [18]. Each category extends the previous one with additional functions and capabilities a social robot should have. The first category of robots are those that are *socially evocative*. This means that these robots induce people to anthropomorphise them. Robots belonging to this category have no other functions then being anthropomorphised as are, for example many toys. The next more complex category is the *social interface*. These robots provide a simple social, often only reactive, behavior by using human-like cues. As an example Breazeal names museum tour guides. These robots pretend to be social because their social competences are only on an interface level. Robots that belong to the next category (*socially receptive*) have more profound social skills. These robots are able to benefit from HRI by learning or by being more perceptive to human signs. The last and most sophisticated category is the *sociable*-robot. These robots have their own goals and motivations and are aware of their social environment. Furthermore, they are capable of pro-actively engaging people in a social manner by asking for help, among other basic functions. This definition focuses mainly on social aspects of robots and makes no claim as to how they can be combined with a technical system. As such, it is a purely sociological definition.

In contrast the definition given by Duffy [44] considers the technical aspects of robots. He introduces a four layer architecture that distinguishes robots on their technical capabilities and originally focused on multi-robot interaction. The first layer is the *physical* one. This describes how the robot is embodied in a physical environment and is therefore physically existing. Duffy then defines the *reactive*-layer. Robots that belong to this second layer are capable of reacting autonomously to their surroundings. They include robots that have sensory perception and actuators such as joints or motors to interact with the environment. The *deliberative*-layer describes the capacity of robots to have their own beliefs, desires and intentions. These three layers are arguable enough to design a full workable functional robot that does not have to cooperate with humans. In terms of

HRI the last layer, the *social*-layer, contains all the capabilities necessary to act in a social manner.

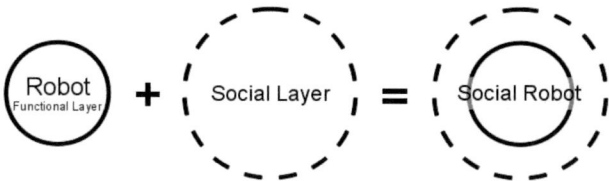

Figure 4.1: The social Layer. A social robot is defined as a functional robot that is surrounded by a social layer. Taken from [69] p. 29

For this thesis Duffy's definition was used mainly for the theoretical and the system part. What follows is a more simplified version inspired by Hegel [69]. This definition distinguishes between two layers (see Figure 4.1). The first is the functional robot itself that becomes a social robot when a social layer is applied. This social layer contains all social features of the robot including its appearance and social behaviors. The functional layer combines the first tree layer defined by Duffy. As mentioned before the aim is to improve a robot by adding more social features. This means that DBCs are part of the social layer that converts a robot to a social robot. Although obviously more is required to make a social robot sociable, as a first step only the DBCs shall be evaluated. The same layer structure is used for the system level (see chapter 7). Here all social behavior is produced within a social layer that is separate from the functional layers of the system and provides social competences.

Based on this four criteria can be formulated that should be covered when designing a social robot. When the goal is to create a complete robotic system it is often not advisable to only consider the Human-Robot Interface when designing social features for the robot. For a complete system it is also important to have a look at the information flow, design and implementation. Not doing so may make it difficult to use the already implemented social features on other robots or to integrate additional functions. A social robot, as well as other robots, consist not only of perceivable functions and appearance. It is a complex system that should also be structured to facilitate maintenance and usability for developers. Because

4. REVIEW OF RELEVANT SOCIAL ROBOTS

of this, the four criteria are divided into two groups: user-centered and developer-centered. User-centered is everything that is perceivable by the user such as appearance and behavior. Criteria belonging to this group are based on the findings described in chapter 2 and 3. Developer-centered criteria are important for the developer and mainly focuses on the implementation and system structure. The criteria are appearance (user-centered), behavior (user-centered), usability (developer-centered) and exchangeability (developer-centered).

Appearance comprises all static signs independently of source (human-like, artificial, ...) or perceptual type. The appearance forms the surface impression and therefore creates the first expectations a human has towards the robot. Consequently, much care should be taken here. The design should be familiar so that it does not scare the human or evokes discomfort. It is not important which source or design concept in general is used. Some robots are very human-like, others are artificial or look like animals. The point is to not evoke misleading expectations. Appearance is the first link to functions and these hints need to be fulfilled. The appearance should in addition fit the task of the robot. For example an education robot should look different from a robot that works in a construction area. Appearance is equivalent to the socially evocative robots in Breazeal's definition and is part of the social layer in Duffy's definition.

All dynamic signs are consolidated under the term *Behavior*. Criteria similar to appearance also apply to dynamic signs. The two types of signs are in close relationship, and affect each other. The main point for the behavior criterion is that the dynamics have to fit the appearance. On the one hand, a behavior has to fulfill the expectations or at least has to not evoke expectation mismatches (neither to appearance nor to another behavior). On the other hand, behavior may not be executable without appropriate appearance characteristics in the robot; hand gestures without hands are not possible. Studies have shown that the behavior must not be predictable in order the robot to look more lively. At the other extreme, a too random behavior can also evoke discomfort. Furthermore it is beneficial to use behavior that people are already familiar with, although this is not strictly necessary. Humans have the capacity to adapt to new situation even if this adaption sometimes require a training process. Per Breazeal's definition the behavior consists of the last three categories and is also part of the social-layer in Duffy's definition.

The third criterion is *usability* and describes the ease with which roboticists can maintain and use the implemented system. This criterion advises implementing the underlying robotic system in a way that is simple for others to use and easy to maintain. This increases the quality of such a system, making it possible to create more and better social features. Usability concerns single components as well as communication between them. This enables the exchange of smaller or larger parts of the system (see next criterion) or facilitates to integrate new features without creating complications. With a focus on social robots the most crucial point is the integration of social features into the functional system. Here different strategies can be used. Some prefer a complete integration where the functional and social features are mainly implemented within the same components which is a more syntactical approach. Another approach creates a semantic division where the social features are produced separately from the functional ones. Following Duffy's definition this criterion addresses the communication between the social and the functional layer, and contains the implementation of all layers.

Exchangeability can be formulated as the fourth criterion. This means that in many cases it is a good idea to implement most parts of the system so that they are exchangeable. It should be possible to replace different behaviors and sensors up to and including the whole robot without changing much of the code. Here a middleware like YARP [105] or XCF [164] is useful. Often robotic systems are optimized for a single robot. This means that different modules , such as a behavior module, are produced with robot specific code. If there is a division between behavior production and command execution (e.g. by using flexible interfaces) it is possible to use the same behavior pattern for different robots. This means that the underlying system does not have to know which robot is used.

In the following sections a deeper look at different social robots is provided. The robots will be presented in more detail, and will be analyzed with a view to the four criteria for social robots.

4.2 Kismet/Leonardo

Leonardo and Kismet are two social robots designed in the Media Lab at MIT. Both were created to conduct social interactions with humans, and are capable of perceiving information from people to produce social feedback.

4. REVIEW OF RELEVANT SOCIAL ROBOTS

Figure 4.2: Kismet (left) and Leonardo (right): Two robots designed at the Media Lab at MIT

The appearance of Kismet [21] is very artificial yet has rich human-like features like eyes, ears and a mouth (see Figure 4.2 (left)) making it possible to express a variety of emotions. The robot itself only consists of a head and it has no actuators like hands or legs. This might be a disadvantage for this robot with respect to appearance. Talking to a head that is mounted on a flat surface may evoke discomfort. On the other hand, the design of the face is very impressive. It is a good abstraction of a human face so that it is easily anthropomorphised while upholding its artificial design. In addition, it has animal-like ears. This abstraction is a good way to evoke lust the right amount of expectation for the behavior of the robot. Leonardo [19] on the other hand, has an animal-like design(see Figure 4.2 (right)). The robot is a relatively small (ca 60 cm tall) humanoid robot with large ears, and a long brown hide covering its body. This design is reminiscent of a toy, and in combination with its size also evokes the right amount of expectation fitting the task of the robot. Leonardo was designed mainly for tutoring tasks where a human teaches the robot something.
On a behavioral level Kismet provides a large number of different social features that perfectly fit the expectations produced by its appearance. According to Breazeal's, this robot can be classified as sociable robot. It has its own goals and motivations and is able to actively perceive and interact with its environment. It is able to show convincing emotions, can recognize affective speech, can produce speech with fitting lip movements and has an attention system. All these behaviors fit the appearance (e.g. speech is expected when the robot has lips) and in addition, the motivational system

makes the robot very lively and has no disturbing behaviors that are new to the human user. For example, it becomes bored and fells asleep if there is nothing happening, or is able to react to a toy if someone wants to play with it. The behavior of Leonardo is not as complex as Kismet's although it is more focused. Leonardo's behavior is reduced to that necessary to improve a tutoring task [20]. Beside simple social background features like blinking, gazing and idle movements this robot uses social functional gestures such as nodding, head shaking, showing confusion and joint attention to improve the tutoring. As such, the behavior and the appearance are also well combined. The most noticeable feature of Leonardo is its large ears. This feature may evoke the expectation that this robot is a good listener, which is perfect for a robot designed to learn in a tutoring situation. For both robots it can be said that their movements are well designed and therefore facilitate anthropomorphism.

The usability of both systems is hard to judge because there are no publications on this topic. Social behavior is fully integrated in the robotic system used by Kismet. This means that in terms of social robotics, the social system and the functional system are not strictly divided. There are modules that compute the motivation and others that compute reactions onto external input and so forth. The behavior itself is mostly created within the behavior module that produces the social and functional behavior in combination. This approach has the advantage that one module has an overview of the whole behavioral system, and is therefore capable of creating an optimized behavior where functional and social movements can be aligned to each other. On the other hand, this approach lacks flexibility. The whole system seems well structured which is important for maintainability. For example the attention system is a separate module as is the motivation system. The robotic system of Leonardo is an extension of the artificial intelligence system C4 [27], with a set of high level capacities added, i.e. goal based decision making, task learning and task collaboration. That aside, much the same can be said for Leonardo as for Kismet. The system consists of a clear structure where different jobs are done by different components.

In terms of exchangeability, it is not known whether the possibility exists for these two robots. The use of dedicated modules for different functions may be a clue that parts of the system can be exchanged although there is no evidence that this might be the case.

4. REVIEW OF RELEVANT SOCIAL ROBOTS

4.3 Nao

Nao is a small humanoid robot that is invented by Aldebaran [62]. This robot is 58 cm tall, has up to 25 DOF and has a variety of integrated sensors, including an accelerometer and cameras.

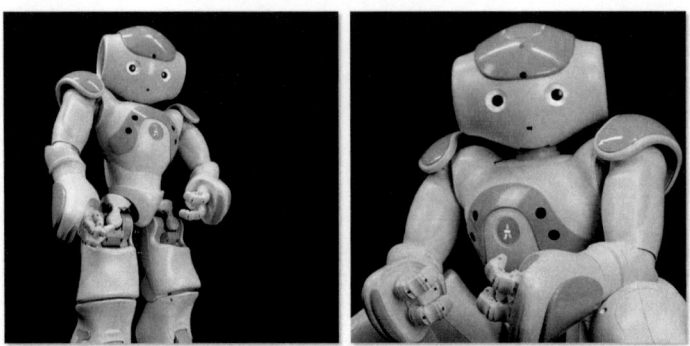

Figure 4.3: Humanoid robot Nao invented by Aldebaran

The appearance of Nao is artificial, although it has a general humanoid shape with two arms, legs and a head. The size of the robot is relatively small for a humanoid robot, making it appear to be toy, a classification supported by it's encasement. The design was well chosen given that the robot was created to play soccer in the RoboCub challenge. In terms of social robotics, the facial features are a bit poor. The whole face is rigid except for the eyes that can change colors using LEDs. The general design of Nao is very child-like with its relatively large head in comparison to the body. This again creates expectations that humans would have towards children which fit the capabilities of the robot. It also fits the task of playing football as younger children love to do the same.

The robot itself has no standard robotic system. The manufacturer provides some high-level features such as walking, standing up and so on but the main behavior are implemented by those who buy the robot. Behavior as a whole cannot be judged because each owner implements its own. The only behaviors that are provided by the manufacturer are some movement demos that show the general capabilities of the robot. In these demos the robot is mostly doing typical children stuff like reenacting movie scenes from Star Wars. This demos show that the robot can be very human-like and familiar when the right movements are carried out, although this depends on the specific implementation of the owner. The missing facial features make it impossible to display emotions directly. Some researchers are working on whole body emotions (e.g.:[66]) to address this deficiency. Due to the missing system, not much can be said about its usability or exchangeability. The manufacturer does provide a profound tool set to integrate the robot into the owner's system structure. With this set, it is possible to program behavior, let the robot walk and to read out the sensors. Even though there is no standard behavior, this robot is useful to researchers interested in testing their own behavioral algorithms

4.4 Geminoid

Geminoids are android robots designed for teleoperating tasks. The Geminoid HI-1 [137] was designed at Osaka University in Japan and is an exact replication of its inventor Hiroshi Ishiguro. Its intended use is as a communication device to interact with students or employees when he is not present. When using this robot, interaction partners are expected to feel the presence of the operator more than with video conferencing. A short time later, an updated version of this robot, the Geminoid DK [29], was presented with the same functions but as a duplicate of another person.

Figure 4.4: Humanoid robot Geminoid: Geminoid HI-1 together with its inventor Hiroshi Ishiguro (left), Geminoid DK (right). Pictures taken from [137] and [29]

4. REVIEW OF RELEVANT SOCIAL ROBOTS

The appearance of the Geminoids is very human-like. The goal in designing this robot was to create an exact copy of a real human being. If the robot and the copied human are standing beside each other it is often difficult to distinguish which one is the real human as seen in Figure 4.4 (left). Compared to the other robots that are presented in this chapter, these robots occur on the human-like side of the uncanny valley. The designers intended to make them as human-like as possible. This strategy has the disadvantage that humans have high expectations towards these robots because of their human-likeness. Further, the appearance of an adult human was chosen, increasing the expectations even further.

The designers of these robots are aware of the expectation problem. This is why they have begun to evaluate solutions to address the issues. They implemented idle behavior such as breathing and blinking to make the robot more human-like, but never evaluated the influences of single cues on people. They also created an interesting approach of generating the movements of the robots [137]. Since they are designed as teleoperation robots, an operator controls most of the robot's movements remotely. For example, the robot is intended to look in the direction the operator is looking , and it can imitate human facial expressions. Idle motions are generated in parallel so that they do not have to be generated by the operator. By using this approach it is easy to generate the right timing for turn taking in an interaction because they are directly generated by a human. This also means that the robot is not autonomous. All these mechanisms lead to a more familiar robot as shown in a study with the Geminoid HI-1 [137] although some movements are not yet quite human-like. These are small technical issues such as arm movements that do not stop the way human arms do. These discrepancies make the robot appear a little bit scary, a typical problem when using robots that nearly look like humans. Humans expect very human-like movements from a very human-like robot. Since this robot is intended for teleoperaton, its usability differs from that of other systems. The main consideration here becomes the usability of the remote control. If the robot is not able to execute the commands of the human user with adequate accuracy and speed the timing and recognizability of the movements can cause negative effects. From experience reports and publications [137], it can be assumed that the remote control of the Geminoid was designed in an intuitive way. The operator simply has to sit in front of the operator desk, and the system classifies the facial expressions and gaze direction, and is able to recognize the lip movements that are then executed in parallel on the robot. Spoken language is also conveyed.

In terms of exchangeability it can only be said that other studies have been carried out on the Geminoid where the robot produces its behavior autonomously. Information on whether middleware or other mechanisms for facilitating exchangeability are used is not available.

4.5 Barthoc Jr./Flobi

Both robots were designed at Bielefeld University and are used to investigate Human-Robot Interaction.

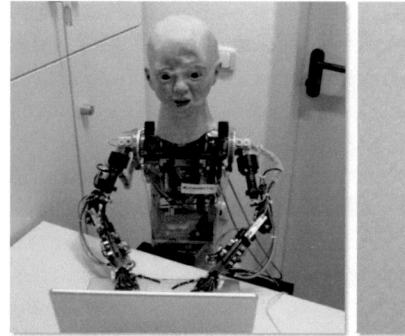

Figure 4.5: Humanoid robots Barthoc Jr. (left) and Flobi (right)

Both robots have a human-like appearance. The Barthoc Jr. is the first version of the robot that was later replaced by the Flobi. The Flobi is composed of a new head mounted on the same torso as the Barthoc Jr. At the moment additional Flobi heads exist that have no body. Barthoc [71] was designed to function as an interaction partner with the human-like appearance of a four year old child. The head of this version was covered with soft skin to mimic real human skin. The head was mounted on an artificial looking body (see Figure 4.5 (left)). As already mentioned for the Geminoid, a human-like appearance evokes high expectations that could not be fulfilled by Barthoc Jr. Its appearance was not accepted and it was difficult to recognize the displayed emotions [70]. Another disturbing aspect was that there exist several mismatches within the design of the

4. REVIEW OF RELEVANT SOCIAL ROBOTS

robot. For example the skin, unlike the skin of a child, had many wrinkles. The eyes and the mouth were too wide open and the human-like head and the artificial body does not fit together. In addition some participants of interaction studies said that the robot could work in a ghost train [95]. These experiences let to a redesign of the head. The Flobi head (see Figure 4.5 (right)) has a more artificial appearance with many human-like features [98]. By using this design, it makes it easier to create familiar robots because of the relatively low expectations compared to a human-like design. Most of the visible materials are some type of plastic which facilitates the association with a toy. In addition the face was designed using a baby face schema. Both let the robot appear cute and friendly, consistent with the goal of designing a four year old child. One of the most interesting features of the Flobi head is that it has exchangeable static cues such as different mouths, eyebrows and hairs. This makes it possible to change the gender of the robot, for example. User studies [70] have found that the emotions of the Flobi head are better recognized compared than with Barthoc Jr.

The Flobi robot, the same as the Nao, has no standard system that produces behavior that can be judged here. The robot is designed to give rich opportunities to integrate it into an external system, and provides an easy to use interface to increase usability. For example, it is possible to use XCF as middleware to send commands to the robots, and there is an XML-Structure that provides an abstraction from the pure motor commands. For example the gaze can be set by sending 2D coordinates to the robot instead of setting the joints of the eyes directly. The standardized interface makes it easy to exchange different components.

4.6 iCub

The iCub [106] is an open source humanoid robot that was built and designed at the Italian Institute of Technology (IIT). The robot is designed so that different researchers can use and develop components for the same hardware. A whole community emerged around the robot that can share and distribute software for all kinds of functions.

The design of this robot consists of two different design concepts as shown in Figure 4.6. In general, its appearance is artificial and it was designed as an infant. On the one hand, the robot has flat, white, plastic skin that is used to convey the social features as the face and arm/leg

Figure 4.6: Open source humanoid robot iCub

covers. On the other hand, the joints and the hands have a technical appearance since the control boards and cables are visible. This design has two advantages. First it facilitates anthropomorphizing the robot because of the human-like static features. Second, the technical appearance does not evoke the expectations of a real human because people see it as a robot. This is why the general level of expectations is not so high making them easier to fulfill. One uncommon feature in the design of the iCub is the facial features. For this robot, the eyebrows and the mouth are realized by using LEDs. It is therefore only possible to display discrete emotions and lip movements. This produces an iconic appearance further decreasing the expectations for facial expressions but also decreasing the possibility for a programmer to display emotions.

The iCub itself is a robotic platform which does not provide a standard behavior, the same as the Nao or Flobi. Over time more and more plug-ins have been programmed by the community to implement different behaviors and functions despite not being standard. For example, a gaze module exists that makes it possible to perform human-like eye-neck coordination by only setting a 3D point to which the robot shifts its gaze. Another module provides human-like arm movements which only needs a 3D point to which the hand should point. However, most of the functions that are produced by these modules are functional movements like crawling, shooting with bow and arrow, or direct joint control. Up to now, only a little amount of social work has been done with the iCub. An example is the work by Katrin Lohan who investigates strategies to better interpret the tutoring behavior of a human to improve the learning task [92]. That said,

4. REVIEW OF RELEVANT SOCIAL ROBOTS

the appearance and behavior that can be produced by this hardware give it a high potential to become a good social robot.

YARP is used for this robot as a communication framework. This is a middleware that makes it possible to communicate with the robot by using a standardized interface, allowing an easy and flexible communication. The variety of modules that can be combined to easily build up a robotic system, shows that the exchangeability of the robot and its components is very good.

4.7 Miscellaneous Systems

In this chapter additional systems will be introduced shortly, where the focus lies on striking features without giving a complete analysis of all four criteria.

Figure 4.7: Miscellaneous social robots: Philips iCat (left), Sony AIBO ERS-7(right)

Another humanoid social robot is KASPAR [10] designed in Hertfordshire. This robot is a child-sized robot with minimal expressive features. This means that the robot has only a few joints (8 DOS for the head) and very few facial features. For example, the robot has no eyebrows. In general a human-like appearance was chosen, including a skin colored face. The choice to use only minimal features was made to decrease the expectations people have toward the robot. The idea for this design was taken from different disciplines, and provides a good review of face design inspired by narrative art design [10].

Another important type of social robots are animal-like robots. This thesis does not focus on this type of robot though they should to be mentioned here because they illustrate important aspects of social robot design. In general, the advantage of using an animal-like robot design is that these robots are compared to a pet and therefore the expectations are different than for a human-like design. The Philips iCat [23] for example, is a robot that is designed like the head of a cat (see Figure 4.7 (left)). The yellow skin gives this robot an artificial comicbook-like appearance. In addition, the robot uses LEDs to show different internal states such as anger and happiness. This robot is designed to function as an interface for humans to control devices like video recorders. The appearance of the iCat fits this task well. By using the appearance of a cat the robot can be seen as a pet. Pets are not unusual in personal homes, and therefore helps the robot fit more easily into daily life. Further, the expectations are relatively low for its capabilities and functions. The use of plastic skin links it to the devices the robot controls since they are also made of plastic. A second example for animal-like robots is the AIBO invented by Sony [54]. This robot was designed as a four legged dog and mainly functions as a toy for children. The appearance of this toy is very artificial and mainly consists of plastic body parts (see Figure 4.7 (right)). Like the iCat the AIBO also uses LEDs as an additional way to communicate by displaying iconic signs on a small screen. The advantages of this design are the same as for the iCat. The expectations are relatively low and the design of a pet makes it a perfect companion for children. In addition, Sony includes complex behavioral programming with the robot which facilitates its liveliness. This behavior consists of idle behavior, walking and simple games that can be played with the robot. All these behaviors are consistent with the dog-like design. It displays the behaviors of real dogs such as recognizing its owner and the need to rest(recharge). The last example is the seal robot Paro [154] which is designed to be used in elderly care. The main focus for this robot is to reduce stress and to give elderly people the task of caring for the robot. To make the robot lively and acceptable, the shape of a young seal with a fluffy white hide was used. This makes it perfect for hugging and stroking. An additional unusual feature of this robot is that its body emits warmth which facilitates the illusion of a living creature. The whole design and behavior evokes the feeling that someone has to care for this robot, which fits perfectly with its function. What these three animal-like robots have in common is that they combine an animal-like appearance with appropriate behavior. This supports the idea that behavior should

4. REVIEW OF RELEVANT SOCIAL ROBOTS

fit expectations. These examples also show that a combination of different appearance types can be used to improve the communication. For example the iCat and AIBO use artificial signs to communicate, and the iCat has a human-like mouth and uses verbal language to fulfill expectations. The combination of different appearance types is possible if they do not evoke expectation disconfirmations.

Beside the robotic systems, software based frameworks have become very important when discussing developer-centered criteria for social robots. These frameworks help to build up a whole robotic system by reusing pre-defined components, and helping to connect all the components that are needed. In most cases, functionalities to debug the system and to observe internal information flow, are available. One famous example for this is ROS (Robot Operating System) [124]. This framework is already used for various robots, and provides rich features for communication and maintainability what fulfills both developer-centered criteria. An example is the Robot Behavior Toolkit [76] developed by Huang et al. using ROS. This system is very similar to the system developed for this thesis but follows the strategy of fully integrated social and functional behavior. Therefore, it does not provide a social interface that allows decoupling of functional and the social system, and the generation of both behaviors independent from each other. Another framework that can be used to build a robotic system is XCF which is used in this thesis and will be described in more detail in section 7.2.

4.8 Implications

As a conclusion for this chapter it can be said that it is not possible to give a concrete guideline for how to build a social robot. The literature shows there are very different strategies for design, and even more solutions to address problems that arise. Still, some strategies emerged that appear promising when designing a robot, and that most of the discussed robots have in common. First, most designers were careful to avoid disturbing features within the appearance of the robot. This refers to misleading features that are not consistent with the rest of the design or function. The second point is that the behavior was mostly designed to fulfill the expectations that emerge from the design itself, at least in the most important cases. There were also examples that extended the expectations and features of the robot in a reasonable way. Thereby it is not so important whether the appearance and behavior of the robot are more human-, animal-, or artificial-like. What is crucial is that the appearance and behavior are consistent. A good foundation for a social robot requires consideration of these two points. In terms of the developer-centered criterion, it can be summarized that a flexible interface for the robot and different modules as well as a reasonable separation of certain functions on an implementatory level is extremely helpful. This makes it possible to exchange functions between researchers that can then use them for their system or robot. Taking all of these points into account appears to be a good strategy for designing social robots. Hopefully this will prevent different researchers from reinventing the wheel because of issues with compatibility or exchangeability.

When designing the robotic system that is used for this thesis We tried to fulfill the four criterion as much as possible and combined ideas from different social robots. As mentioned before, the iCub is used here because of the following reasons. The appearance of this robot is very consistent and has a friendly and cute design (see chapter 4.6). Because the goal is to investigate effects during Human-Robot Interaction for cooperative tasks this appearance seemed to be well-fitting. The character of the robot is friendly and helpful. For behavior, we focus on examining the effects of DBCs on the human. For that purpose, five behavior were chosen that were most salient in HHI. In Table 4.1 a list of exemplary background behaviors is given that was used to build the final set of DBCs. This was gathered by previous work of other roboticists and by observing HHI. Several constraints conducted the decision which ones to choose. On the one hand, limitations in the hardware of the iCub prevented the use of certain behaviors. On the other hand, some were not observable during a cooperative face-to-face interaction. During selection process and design care has been taken that the DBCs fit the appearance of the robot, and that there are no important expectations left unfulfilled. Finally, the following DBCs were chosen: blinking DBC, natural movements, idle movements, gaze DBC and lip DBC. The entire investigation about how DBCs are designed, parameterized and evaluated, and what effects they have within HRI is presented in chapters 5,6 and 8.

In terms of system structure, the implementation is done in such a way that all components are easy to use and to exchange. Also, it is structured so that distinct functions are realized by separate modules as advised by the fourth criterion. In addition, some modules to better control the iCub were used that were implemented by the iCub community which fulfilled the same criterion. A layer structure, although uncommon, was used as a

4. REVIEW OF RELEVANT SOCIAL ROBOTS

Background behavior	For example used in	Included in
Blinking	Geminoid	Blinking DBC
Breathing	Geminoid	Idle movements
Gazing behavior	-	Gaze DBC
Human like gaze	iCub, Geminoid, Kismet	Natural Movements
Human like movements	-	Natural Movements
Idle movements	-	Idle movements
Lip movement	Kismet	Lip DBC
Movement of ears	Kismet, Leonardo	-
Muscle contraction	-	-
Warmth of skin	PARO	-

Table 4.1: List of exemplary background behaviors in HRI and their use as DBC

general structure. Here the social features are produced separately from the functional system. This makes it possible to develop them separately, and to provide a better abstraction that wraps the social robot as one separate system that has an interface to the functional section. This was done so that the developer who creates the functional behavior does not have to consider the social behavior and vice versa. This is comparable to the model-view-controller architecture [26] from software engineering where the data does not have to know in which way it is displayed, i.e. in a Graphical User Interface (GUI). Both are strictly separated to improve exchangeability, flexibility and robustness. In addition, a separate tool was developed that makes it easy to create movements like behavior or whole body movements for the robot without having to program them. With this tool it is possible to configure complex joint movements in an intuitive way by using a Graphical User Interface. The layer structure is a major advantage for exchangeability. The functional layer, the social layer and the robot have no direct dependencies and can be exchanged for other implementations or systems. This is possible because only patterns are generated that are interpreted by the robot. The robot only perceives the command to "nod" and the robot knows which joints have to be used to execute this command. The same command sent by the same system can be executed on a different robot when it knows how to execute a nod. A detailed description of this system architecture and its implementation is given in chapter 7.

Compared to other social robots, my approach is not to create one social behavior that is executed on one robot but to create parameterized behavior patterns that can be used on different robots, and which can evoke different effects when changing the parametrization. Furthermore, single behaviors are examined separately from their effects within HRI. We investigated what influence all behaviors (DBCs) have in combination with the familiarity of the robot, and stress level of the participant. In addition, the system architecture used makes it easy to test and share these behaviors with other robots or researchers.

Chapter 5
Proof of Concept

This chapter describes the studies that were carried out to perform a proof of concept. In the last chapters most of the theoretical background was presented as well as the hypotheses and assumptions about how DBCs effect Human-Robot Interaction. The idea tested was whether anthropomorphization of social robots, and the effect of social movement on HRI could serve as the basis for future research. This proof of concept was necessary because it cannot be assumed that people will anthropomorphise the robot iCub without evidence.

The first goal was to test if the assumptions presented in the previous chapters were true for the iCub. As discussed in chapter 3, it is necessary to test each robot's general design for acceptability as a social robot, and if it is anthropomorphised as awaited. Expectation mismatches within the design or behavior will decrease the familiarity. These first tests were intended to identify these mismatches and and in which way the DBCs used could be improved.

The second goal was to test the first hypothesis presented in section 3.3.

Hypothesis 1. *Increasing human-likeness by using DBCs increases familiarity and perceived liveliness of the robot within a cooperative task.*

It proposes that if the robot uses Dynamic Background Cues, its perceived familiarity or human-likeness will be higher than if the robot only shows functional behavior. To test the hypothesis findings by Heerink et al. were used [68]. The authors argue that increased social expressiveness by the human user refers to a higher state of comfort in HRI, and is therefore a subjective measure. Furthermore Heerink et al. found that a more expressive robot increases comfort [68] which is in line with the first hypothesis since DBCs make the robot more expressive.

Two studies were carried out where participants directly interacted with

5. PROOF OF CONCEPT

the robot in a tutoring situation. The studies used video data and questionnaires to investigate what effects took place, and if there were any disturbing signs that provoke expectation mismatches for this robot. These outcomes will be used to improve the signs that are produced by the iCub. To generate all behavior, a preliminary version of the robotic system introduced in chapter 7 was used.

5.1 Social behavior

To create a social condition used in this study a social behavior was built that uses mainly the set of salient DBCs introduced in section 4.8. This behavior enriches the functional behavior used for the non-social condition by applying DBCs. Since being a proof of concept the used DBCs are preliminary versions which were improved upon the findings of this study. The improved version will be discussed in more detail in chapter 6. Further, only three DBCs are used. The passive role of the robot in this scenario did not allow to generate lip movements and the gaze DBC was invented upon the experiences gathered in this study. Therefore, only the blinking DBC, natural motion and idle motion (breathing) were used.

The blinking DBC was an adaptation of that used by humans, who normally blink at a rate of 12.55 blinks/min. Similarly, the robot was programmed to blink once every 4-5 seconds. The purpose of including breathing behavior is to perform some idle movements to make the robot more lively. These movements allow the robot be in motion all the time and to never appear stiff. The movements are therefore, very subtle and slow. Because of limitations in the robot's movement abilities, chest expansion was simulated by having it lean back and forth slightly. Combined with the synchronized arm movement, this resulted in a convincing breathing animation. Natural eye-neck coordination was implemented using the `iKinGaze` module programmed by the iCub community. This provided human-like motion by using saccades with the eyes fixated on the destination point while moving the head a little bit slower in the same direction.

5.2 Study design

Both studies were identical in terms of design. The second one was done because we wanted to have a more detailed questionnaire that focuses more on social topics that were not covered in the first one. Evaluation of the video date was done in the first study. Because of the identical setup of both studies only one study setup will be described.

The scenario was a tutoring situation where the participants were asked to teach the robot different tasks. For example, the robot was taught how to switch on a lamp or how to stack 4 cups together (see Figure 5.1). During the study, the robot and the participant sat facing each other at a table. The robot in this study was the recipient of tutoring by the human. The participants were told to explain the tasks in a way that they thought would be natural for them. These instructions led to different teaching strategies. Some participants showed the robot the movements without speech, some used technical language, and others used a didactic style suited to an infant. This variety of teaching styles was intended because it is hypothesized that the participants act in a more social way depending on the amount of social behavior expressed by the robot. According to Herrink, this may be a clue to increased comfort when interacting with the robot. This scenario was chosen because it is a cooperative task requiring social interaction. Letting the human participants teach something to the robot puts the robot in the passive role of a listener while putting the participant into the active role of the teacher/parent what further facilitated the perception of the robot as an infant. The teaching scenario has advantages to evaluate social effects because the human can directly address the robot, and the behavior toward a robot reflects how the robot is perceived. The tasks here were previously used in similar human-infant studies [132].

Two conditions were created for the studies. The first condition was the *non-social* condition where the robot performs functional behavior by only moving the eyes. The second condition was the *social* condition where the robot uses social behavior. Each participant only saw one of the two conditions (between-subject design). During the study the robot could only be seen from the hip upwards because of the table. As a reactive behavior, a salience system modeled after Itti and Koch [78] was running on the robot. The robot's visual perception was recorded using a webcam mounted above the iCub's head, and then analyzed for salient regions using features like color and motion. This gave the robot the ability to look in the direction of the most salient salient regions such as faces or moving objects. In this study salient regions such as colorful objects in motion, or the participant explaining something, such that the face becomes salient because of the movements of the mouth, emerged. In some situations this behavior might appear random because the most salient point is sometimes

5. PROOF OF CONCEPT

not something the participant expected the robot to look at. More details will be covered in the discussion section. This gazing behavior was chosen as a functional behavior since it established joint attention between the robot and the subject which was then often used in a cooperative task as well as in both conditions.

The non-social condition was very simple. Only gazing behavior was turned on, with the robot performing the gazing only with the eyes. The rest of the robot was completely at rest. In the social condition the robot performed the same functional behavior as in the non-social condition but in addition also used various DBCs as background behavior to appear more social and familiar. The detailed social behavior was described above.

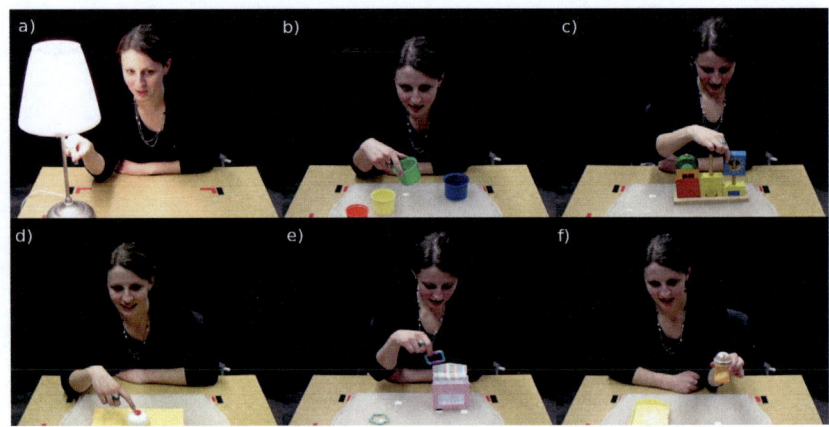

Figure 5.1: Participant performing six tutoring tasks: (a) switching on a lamp, (b) stacking four cups together, (c) building a little toy-house, (d) ringing a bell, (e) placing rings in a box and (f) using a saltshaker

The study itself was carried out as follows. First, when the participants entered the room, the robot was already turned on and performed the behaviors depending on the condition the participants were in. It is advisable to provide the appropriate expectations of the movements from the beginning when the robot is seen for the first time. After that, the participants were asked to sit down in front of the robot. They were asked to describe six small objects to the robot in a way they thought was natural and adequate. The experimenter then fetched some objects, put them in front of the participants, and gave instructions such as: "Show the robot how to

switch on a lamp." Then the experimenter vanished behind a wall to not disturb the participants. The participants proceeded to show the robot this single task while the robot reacted depending on the condition it was in. The object was then exchanged for a new one. All instructions were given in German because the whole study was carried out in Germany. All instructions have been translated to English in this chapter. The six tasks were adapted from Rohlfing et al. [132] and can also be seen in Figure 5.1:
Switching on a lamp (a): To perform this task the participants got a little table lamp and was told to teach the robot how to switch it on. For this task, the lamp was connected to a power supply so that it can be lit by pulling the attached cord.
Cup stacking (b): For this task the participants got four nesting little cups in different colors. The blue one is the biggest followed by the green, yellow and red. The instruction for this task was to show iCub how to stack all cups into the blue one.
Building a house (c): For this task the participants received a set of toy blocks partially build into a house. The participants had to show iCub how to position the remaining three blocks so that the house looks like the one in a small picture available only to the participant.
Ringing a bell (d): Here the participants were given a bell, and asked to show the robot how to ring it. The bell was fully functional so that it could emit a sound.
Placing rings into a box (e): For this task, participants were given a little box with three rings lying beside it. The box was closed at the beginning. The participants were told to show iCub how to put these rings into the box.
Using a Saltshaker (f): The last task was to show the robot how to use a saltshaker. A filled saltshaker was given to the participants, along with a small bowl for salt to be shaken into during the demonstration.
For all tasks, the objects (except the lamp) were placed on identical trays that were put in the same place. This ensured that the objects have the same position for all participants. To prevent ordering effects, the tasks were presented randomly. The only exception was the lamp task. It was always the first item and was intended as a practice task.

The setup used is shown in Figure 5.2. The participant sat on the right side on the chair. The iCub was placed on the opposite side. All objects that were necessary for the task were placed in the middle of the table. Three cameras were used to observe the scene and to collect data. Two cameras, one behind (cam 2) and the other in front (cam 1) of the

5. PROOF OF CONCEPT

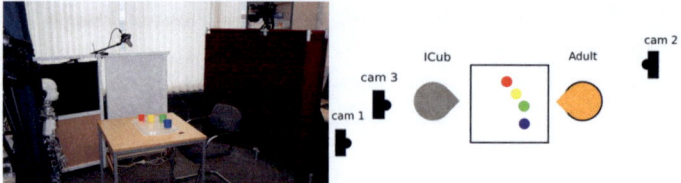

Figure 5.2: Study setup

participant were used for data recording. They were arranged so that both the behavior of the iCub as well as the behavior of the participant was recorded for further evaluation. In addition, a webcam (cam 3) was placed above the head of the iCub which was used to gather data for the salience module to produce the gazing behavior during the study. Two visual covers were used to occlude distracors. A curtain was placed behind the iCub to cover the power supply and camera 1 and 3. A movable wall behind the participant was used to cover camera 2 and the operator desk. Furthermore the experimenter used this cover to hide after presenting the objects. All this was done to provide a familiar environment for the participants by camouflaging a possibly uncomfortable and distracting lab environment. One success is measured by the face that most participants reported that they did not recognize the webcam above the head of the iCub despite being in their direct field of view.

5.3 Measures

To evaluate these studies, three measurements were used. In the first study, the quantitative video data were evaluated with respect to the behavior of the participants. During the second study questionnaire data were gathered to evaluate the perception of the robot by the user. Qualitative user feedback was collected during both studies to get an impression of the robots signs that needed improvement.

The quantitative video data were coded as follows. Three feedback types were used to classify user feedback. For each type,each occurrence was counted during the interaction, and these values were compared depending on the condition the robot was in. The first type is *directly addressing* the robot. This means that we counted how often the participants address the robot directly. For example by saying "Hello iCub", "Can you see this?" or "Shall we do this together?". Directly address is a social action by a human that shows that the robot is seen as a social entity. On the other hand, the robot could be treated as a kind of tool where less direct addressing takes place and the tasks are explained in a more technical way. The second type was the *direct attention*. Participants that tried to direct the attention of the robot actively tried to focus the attention of the robot on a certain object. This could for example be by holding the object near the robot, by shaking the object or both. Such behavior can often be observed when parents teach their children similar tasks, and is called motionese. More details about motionese and their impact for Human-Infant Interaction can be found in [132] and [14]. This behavior also signals the attempt to behave socially since this behavior is used in Human-Infant Interaction. Its presence is evidence that human participants ascribe infant traits to the robot. *Pointing* is a functional measure. These gestures occur in social as well as in non-social contexts but only transfers less background information. We only counted how often the participant points somewhere and ignored the background information. By using these three measurements it is possible to evaluate if DBCs have an influence on the social as well as on the functional behavior of the participant.

Two different item sets were used in the questionnaire. The first set was a collection of items referring to which human traits are ascribed to the robot, and what skills the participants believe the robot has. The general questionnaire consists of 19 items in German that were inspired by items from different robot interaction studies. The original German version can be found in the Appendix Figure A.3 as well as a translated English version in Appendix Table B.3. All items use a 7 point scale. Sample items include, "Do you think you and iCub have much in common", "Do you think iCub understood the taught objects", "Did you had fun during the interaction with iCub" or "How human-like are the movements of iCub". Responses were logged using a 7 point scale, with 1 indicating lowest agreement and 7 highest. The second set was the Big5 scale [79] which contains 10 items that refer to 5 character traits (openness, conscientiousness, extraversion, agreeableness and neuroticism). The original questionnaire can be found in the Appendix Figure A.2. Each item uses a 5 point scale ranging from "Strongly Disagree" to "Strongly Agree". Each of the five dimensions contains two items. Both refer to one extreme of their corresponding dimension (eg.: extroverted and introverted). To compute a score for one dimension, the value of both items is summed up

5. PROOF OF CONCEPT

($extraversion\_score = introverted\_item + extroverted\_item$). The values of both items are scored using mirror scales, i.e. for the first item "Strongly Disagree" is scored with 5 and "Strongly Agree" with 1 for the second item "Strongly Disagree" is rated with 1 and "Strongly Agree" with 5. For example, to get a high score for extraversion, introverted must be strongly disagreed (high score) and extroverted must be strongly agreed (high score). The Big5 was originally intended as a self-rating scale, although in this study it was used to rate the character of the robot. Further, the questionnaire contained items concerning age, education, computer knowledge and another Big5 that were directed towards the participant. The questionnaire began with an introduction each participant had to read which instructed them to answer intuitively and without thinking too much. The original version of the introduction can be found in the Appendix Figure A.1

The questionnaire also included qualitative items where the participants were instructed to answer in their own words. For the first study participants were asked "Do you think the robot had understood the demonstrations?". The second study had a follow up question where participants were asked to argue why or why not they thought the robot had understood them.

5.4 Results

The results for both studies are as follows. In the first study 18 participants (12 female, 6 male) participated with an age ranging from 20 to 64 years old ($M = 33.64, SD = 12.406$). Nine participants were evaluated in the social and in the non-social conditions. The video data were evaluated for the first study. Four measurements were used to test on differences between the conditions. The first three were the measures presented in the last section (direct address, direct attention, and pointing). As a fourth test, the sum of direct address and direct attention was used to represent the total amount of social feedback. T-Tests were used unless otherwise indicated. Since the data for the direct address were not normally distributed (Shapiro-Wilk Test p < 0.005) a Mann-Whitney Test was carried out. For the **direct address** measure there were significant differences between the non-social ($Mdn = 2$) and the social ($Mdn = 10$) condition, $U = 17.5, p = .038, r = 0.49$. Furthermore there were no significant differences for **direct attention** between the non-social ($M = 5.67, SD = 4.664$) and the social ($M = 9.11, SD = 4.014$) condition,

$t(16) = 1.679, p = .113, d = 0.791$, but the mean values show that in the social condition the amount of attention directing was higher. **Pointing** also had no significant differences between non-social ($M = 1.89, SD = 2.147$) and social ($M = 2.89, SD = 4.076$), $t(16) = 0.651, p = .524, d = 0.307$. For the combined score of direct address and direct attention there was a significant difference between non-social ($M = 9.22, SD = 8.258$) and social ($M = 19.56, SD = 11.631$), $t(16) = 2.173, p = .045, d = 1.025$.

The questionnaires were evaluated for the second study. In this 21 participants (9 female, 12 male) participated with an age ranging from 15 to 63 ($M = 29.05, SD = 12.631$). The questionnaires contained 19 items in total. Here only the most noteworthy items will be presented. A complete list of results can be found in Appendix B.1. To evaluate the general questionnaires, a Mann-Whitney Test was carried out. Participants rated the movements of the robot as significantly more **human-like** in the social ($Mdn = 5$) than in the non-social ($Mdn = 3$) condition, $U = 20.5, p = .013, r = 0.542$. For the second set T-Tests were carried out to evaluate the combined scores of the Big5. The character of the robot was rated as significantly more **open** in the social (($M = 6.455, SD = 0.687$) than in the non-social ($M = 5.5, SD = 1.179$) condition,$t(19) = -2.238, p = .033, d = 0.989$.

Participants had the possibility to give feedback about the behavior of the iCub in their own words for both studies. When comparing responses, two topics occur frequently. These were gazing and feedback given by iCub. For gazing most participants expressed concern that gazing appeared to be random and that there was no direct gazing to the participant's face. On the other hand some participants stated that the robot follows their movements. Overall there were 8 positive and 16 negative comments concerning gaze, although there was no difference between the non-social and the social condition. The second topic concerned the missing feedback that was expected by the user. Many participants said that there was no reaction, and no interaction. Most of these comments did not specify what feedback exactly was missing, although some mentioned missing speech of the robot. There where no positive comments concerning feedback, not surprising giving that in this study the robot did not give any. There was one negative comment for feedback in the non-social condition and 8 in the social condition.

5. PROOF OF CONCEPT

5.5 Discussion

In these two studies we investigated if DBCs influenced the perception of the robot, and in which way the DBCs could be improved. The first study showed that humans directly address the iCub more in the social condition and that the combined score for social feedback is also significant higher than in the non-social condition. This shows that even subtle cues like breathing and eye-neck coordination can influence the human participants to use more social signals himself. According to Heerink [68], this shows that the human is in a higher comfort state when interacting with a social robot. In addition, it should be noted that there are no significant results for the pointing gesture. The reason might be that pure pointing itself is not a social sign. This could mean that DBCs that serve as background cues only effect the social signs of a human participant without effecting the functional ones. Because of this, subsequent studies are more focused on this topic to evaluate this effect in more detail.

The questionnaires for the second study showed that the movements of the robot are significantly more human-like in the social condition than in the non-social condition. Furthermore the character of the robot is seen as more open-minded. Both results support the findings of the first study. When using DBCs the robots is perceived as more human-like and humans ascribe more human traits to the system. The fact that more positive character traits are ascribed shows that the usage of these particular DBCs can be used to create a friendly robot.

In terms of improving the DBCs, user feedback revealed some beneficial results. The lack of robot feedback was the most important criticism from the participants. The fact that the missing robot feedback was especially marked in the social condition shows that these expectations are strongest when the robot behaves in a social manner. The comments concerning the gaze of the robot showed two things. On the one hand, they revealed some problems with the software used to control the eyes. Parameters are used that did not fit the expectations of the user. Gaze was too slow, and was shaky when a fixed region was indicated. On the other hand, the gazing strategy was not adequate for this scenario. The participants expected to be looked at by the robot when interacting with it. This feedback shows that there need to be some changes in the behavior of the robot. First the robot has to produce more feedback when interacting with a human, especially when using DBCs. Second, the parameters of the gazing behavior must be optimized to follow more human-like patterns, and in addition, the gazing strategy has to be changed. The saliency model is a useful approach when interesting points need to be spotted. However this model does not distinguish between faces and objects. In a social interaction, it is often important to face the human instead of gazing at the most interesting object. Our saliency model is not able to achieve this. The idle behaviors of the robot (breathing and blinking) seemed to be accepted by the user; there were no negative comments on these movements. Nevertheless these seem to be important for the behavior of the robot. Beside gaze these two DBCs were the only difference between the two conditions and although gaze was criticized, the overall movements were rated as more human-like. A detailed description of all DBCs and how these comments are used to improve them is presented in chapter 6.

These results show that Hypothesis 1 can be assumed. DBCs are able to make a robot seem more human-like, and can help improve familiarity even though they are only subtle background cues. Furthermore, these studies show that the iCub can be used as a social robot within HRI. The robot is anthropomorphised and the results show that more human traits are ascribed to the robot when it is exhibits more social behavior. Unfortunately, some of the evaluated items did not give convincing results, especially those that refer to social perception of the robot. We hope that by improving the DBCs, and by considering the feedback of the participants, it will be possible to have stronger effects and more meaningful results.

This study has shown that DBCs are able to improve Human-Robot Interaction, and revealed the interesting effect that background signs seem to effect mainly the social aspects of human participant without affecting the functional ones. The rest of this thesis will more focus on examining these effects.

Chapter 6
Dynamic Background Cue Pattern

In this chapter all Dynamic Background Cues will be explained that are used to improve interaction based on the third criteria of social robots. As described in the theoretical background chapter these DBCs are used to evaluate their potential to improve HRI by improving robot familiarity and task performance. Furthermore these DBCs are used to generate robot motion that fits the expectations evoked by the appearance of the robot, as described in the first two criteria for social robots. This chapter will describe the patterns used to realize and generate the DBCs. Some patterns were based on literature, some were improved upon using the results gathered from the studies described in chapter 5. For the remainder online studies have been carried out to evaluate which parameter for these patterns are optimal for the desired task. Furthermore the theoretical background and related work that focuses on HRI will be presented for each DBC. We will also evaluate within the two online studies whether the second Hypothesis is correct. It states that:

Hypothesis 2. *The use of parameterized DBCs makes it possible to create certain robotic character traits that are perceivable by human users both when using single and multiple cues in combination.*

With the following online studies will evaluate single cues exclusively. Combined cues will be examined in section 8.1.

DBCs are often simple and subtle behavior patterns used to enrich the functional movements of a robot to that it can appear more social. By definition they are dynamic cues and they are background cues that have no functional effect on the task. DBCs can be human-like as well as artificial and can affect the human through all possible senses. A more detailed definition of DBCs is given in section 3.2. In this chapter, the five DBCs

6. DYNAMIC BACKGROUND CUE PATTERN

are discussed in more detail that form the set of salient DBCs described in section 4.8: Blinking, natural movements, idle movements, gaze and lip-movements.

Most of the DBCs used are designed as parameterized patterns independent of the robot. This means that they define manner and timing of the behavior without having to be implemented for a specific robot. This makes it very easy to use the benefit that each DBC provides on different robots. For example, if the eyes of a robot should be closed the pattern sends a `close` command instead of setting the direct joint values. The module that realizes the communication with the robot has to interpret these high-level commands. Using this approach allows us to use the same pattern on different robots. The only requirement is that the robot is able to understand the high-level commands. The advantage of using parameterization is that the pattern can be modified when it is needed. For example, blinking behavior has three parameter: The time the eyes are closed, the mean time between two blinks and a variation factor that randomly varies the time between two blinks to make the blinking more unpredictable. By modifying the parameter, it is hypothesized that different characters of the robot can be produced. Possibly a higher blinking rate may appear as more nervous. For gazing and lip-movements, online studies have been carried out to find the optimal parameters that create an adequate character for HRI. In this context, it should be mentioned that these online studies have been carried out with native Germans. Because of cultural differences, it is possible that for example, in Japan, other parameters would produce the optimal desired effects.

The idea to use certain patterns to create a more lively character is not new. In animation movies this topic is essential. For example in the book "The illusion of life: Disney animation" [148] animators at Disney present their approach to creating lively characters. The twelve principles of animation are presented focusing on the animation of a single character. For example, animations that stop immediately or two animations that do not have a smooth transition may appear stiff or unnatural (Principle 5.: Follow Through and overlapping Action. [148] p.59). Many of these principles can be transferred to robots. Breemen gives a good review [22] of these principles and their usage in robots. These principles are excellent guidelines when creating dynamic movements, as such some of the presented DBCs use them to appear more natural.

The structure of this chapter may evoke the impression that all cues can be seen as separate aspects within HRI, with their own requirements and effects on the human participant. It is incorrect to assume that one sign has one effect that can be turned on when the sign is created by the robot. In section 2.2 we discussed that there is a complex interplay between all signs, and that one sign can evoke expectations for another sign. Activating one sign that makes the robot more friendly (like a friendly smiling mouth) raises the expectation for friendly looking eyes. In general, one should be careful that the presented signs fit together. This admission makes it difficult to create a list of effects that each DBCs has on the human user because any given effect can only evolve full efficiency when the other signs support it. If the second hypothesis is correct it can be seen more as that a general character for the robot should be chosen beforehand, and all cues must be tuned in such a way that they fit this chosen character. The character that should be created for our social robot is a friendly cooperative, fun, interaction partner. In the following sections care is given in the selection of parameters that fit the desired robot personality.

As described in section 3.2, which signs are classified as background signs and which as functional signs depends on the robot's task. For the definition of these DBCs, it is assumed that the human and the robot are in a general Human-Robot Interaction which has the goal to solve a cooperative task. This means functional signs are those signs that are directly relevant solving the task, such as giving instructions, doing object manipulation and so forth. The most social signs, such as the DBCs describes here, are not in direct focus of attention within this task and therefore background signs.

6.1 Blinking

The blinking DBC is a simple adaptation of human blinking behavior. It encompasses a short closing of the eyes after a randomly chosen waiting time between two blinks. This behavior is biological inspired by the behavior that is produced by humans in timing and in movement pattern. In biology, blinking is a behavior necessary to suppling the eyes with moisture. The production of this cue is not socially driven and is produced unintentionally. However, its absence may make a robot appear unfamiliar,given that direct interaction involves gazing, and as such contains a strong focus on the eyes. Furthermore, the goal of this behavior is to make

6. DYNAMIC BACKGROUND CUE PATTERN

the eyes more vivid, and to reduce the impression of staring. Humans in general perform 12.55 blinks/min [31] which varies depending on the circumstances such as dry eyes and so on. In addition, humans normally blink while doing saccadic eye movements [50] because the wind produced by moving the eyes so fast triggers the reflex to close the eyes.

The blinking DBC is a human-like cue that mainly influences the human participant visually. The auditive sense is also typically influenced in conjunction. In some robots, such as the iCub, closing the eyes in a speed necessary to perform a blink, produces sounds in the joint motors. To date, there are no studies on blinking behavior in social robots. That said, the study presented in the previous chapter found that participants did not having negative association with blinking. Furthermore, in combination with breathing, this behavior can help a robot appear more human-like. Future research should focus on what effect the isolated blinking cue has.

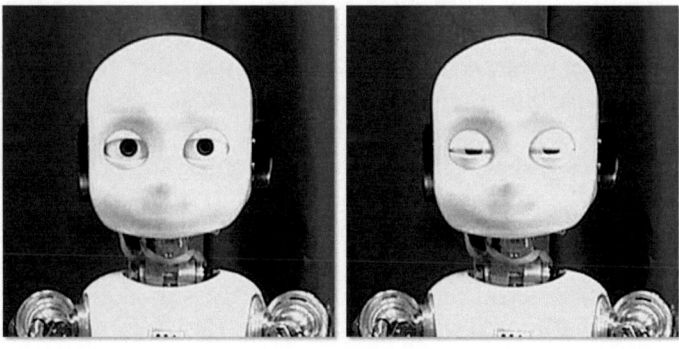

Figure 6.1: Different stages of iCub performing a blink

The pattern used contains three parameters. The first specifies the mean time in milliseconds between two blinks. The second adds a random variation to this mean value. This value determines what percent of the initial time between blinks should be maximally added or subtracted. The last parameter is the time the eyes are closed when blinking. Since the intention was to mimic human blinking patterns the following parameter were used for this DBC. The time between two blinks was set to 5000 ms with a variation factor of 0.3. The closing time for one blink was 100 ms. This resulted in a mean frequency of 11.76 blinks/min. Blinking while performing a saccade was not implemented at that time.

Known problems for this behavior include the previously mentioned sounds produced when the eye-lids are moving very fast. As the feedback within the first study shows, there is no obvious disturbance for the participants. It should be investigated later if the blinking noise produces an unconscious disturbance. In the meantime, it is impossible to eliminate the noise without changing the hardware. Another problem is that blinking occludes the video stream present in the eyes rendering the visual system virtually blind.

6.2 Natural Movements

The second DBC is the natural movement of the robot. This DBC differs from the others because it is not directly parameterized, and describes a more general design for creating robot movements. To create movements that fit the appearance of the robot, it is important that these are "natural","adequate" and "develop over time". These three terms were defined by Saerbeck et al. [135] who developed design guidelines and tools to create believable motion in personal robots. *Naturalness* means that movements should be natural to the design of the robot. This does not mean that a movement need to be human-like. If a robot has a very robot-like design the movements should be more robot-like for example by using only functional movements with fewer different joints and the same velocity. If a robot is more human-like, different trajectories should to be used with more complex motion involving more joints and different velocities. An example is presented in Figure 6.2. Here a pointing gesture is shown. The robot-like movement only uses one joint to raise the arm. In the human-like movement, the elbow is bent very early in the movement, and unbent at the end. Furthermore, the hand follows the arm movements in a natural way. This movement is much more dynamic and is inspired by human motion even though the end position is the same for both. *Adequateness* in Saerbeck's definition refers to the "right moment" in which the motion is to be presented. The "right moment" refers to timing and context of the action. Context is determined by the communication partners observing the robot, their social relationship and by the application scenario."([135] chapter III.B). *Development over time* refers to the variation of similar movements. When designing motion for a robot, repetition of the exact same movements within a short period of time should be avoided. Humans are very sensitive to recognizing biological motion, and no two movements by a biological creature look exactly the same.

6. DYNAMIC BACKGROUND CUE PATTERN

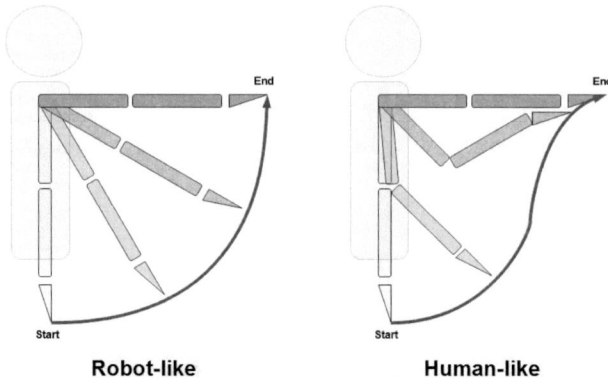

Figure 6.2: Pointing gesture: Performed in a robot-like and in a human-like fashion

The open question is how to design the right natural movements. There are different ways how to collect natural movements for a humanoid robot. On the one hand, it is advisable to analyze humans to adopt their behavior, movements and timing. This can be either be done by observing them directly or by using motion capturing techniques. On the other hand, there is a great deal of literature for creating lifelike motion in animation movies which can be used as a guideline for robot movements. One of the most prominent guides is "The illusion of life: Disney animation" [148] which presents the twelve principles of animation. The following four principles can also be used in robotics.

Principle 2: Anticipation says that to understand a scene there has to be a planned sequence of actions that lead clearly from one action to another. For example, if a character wants to throw a ball, it has to reach back before doing so. Without this movement, the throwing animation appears unnatural because an essential part is missing. In addition, seeing that the character reaches back lets the observer foresee what comes next.

Principle 5: Follow Through and Overlapping Action refers to the connection of two actions. When one action is completed there must not be a full stop. This appears stiff and unnatural. There have to be follow through actions so that it is not noticeable when the first stops and the next begins.

Principle 6: Slow In and Slow Out describes how the animations of one character are carried out in different velocities. The speed is not constant so as to avoid abrupt acceleration and stopping. Movements speed up at the beginning and slow down at the end.

Principle 7: Arcs. This principle states that movements of most living organisms follow arcs, i.e. the swinging of an arm, running feet, and so forth. On the contrary, moving in straight lines emulates machines, such as a car driving by. Using straight lines for the position of an endeffector of an robot arm would make it appear very technical. It is advisable to instead use arcs to support a human-like appearance.

Another example is "The Animator's Survival Kit" [163] by Richard Williams. This book mainly focuses on how animations should look to appear vivid. For example, the human-like arm movement in Figure 6.2 is inspired by this book (p. 232) where the author describes how arm movements without "breaking" the arm appear stiff and rigid. "Breaking" in this sense means that the arm is bent during the movement instead of using only the shoulder joint, resulting in an s-shaped hand trajectory.

The relevance of using these DBCs is very high. As can be seen in the literature, the natural and adequate movement of one character is essential to creating a vivid character the observer can sympathize with. Within this thesis care was taken to design all movements of the robot following these principles. This was done either by designing them manually or by using preprogrammed modules such as the human-like eye-neck coordination (see below). The DBCs are mainly human-like visual cues but, as with blinking, motor sounds can occur when the speed of the movements reaches a certain speed.

Within the social robot community, some work has been done that concerns natural movements. For example Gielniak et al. [57] developed an autonomous algorithm that enables a robot to generate anticipatory motion variants that help the user to understand the goal of the robot's motion earlier. This enables robot handoffs to be more fluid because the user has more time to prepare. In terms of eye-neck coordination, Miani et al. [101] investigated how to adopt human eye-neck coordination to a robotic anthropomorphic head. The goal was to execute fast gaze shifts towards a visual target. To achieve this, a biologically inspired model was used for sensory-motor coordination. Lopes et al. [96] describe a method to control the eyes and neck of a robotic head based on the human oculomotor system during target tracking. The general goal in using natural movements for robots is to make them more lively and familiar. It is hypothesized that

6. DYNAMIC BACKGROUND CUE PATTERN

there are fewer expectation mismatches in the design of the robot if it uses motion that is natural for a human-like looking entity. If this is true stress could be reduced during interaction as human participants can better anticipate movements and decrease uncertainty about the robots movements.

Figure 6.3: Transferring human motion to a robot using video data

Two different types of DBCs were used as a source of natural movement for the robot behavior design used for these studies. There were general movements of the arm and legs, and eye-neck coordination used to perform gaze shifts in a human-like fashion. The general movements used were mainly designed manually. It has already been mentioned that these DBCs are the only ones not parameterized. The reason for this is that the different movements are very different, and no general pattern could be created to cover all possible variations within an interaction. To generate natural movements for the interaction study, without the option of evaluating some parameters, we conducted pre-recordings of four people performing the same interaction the robot would carry out later. The robot behavior and the timing were generated based on this video material. Figure 6.3 shows a snapshot of the human interaction and the corresponding robot movement. The complete interaction as recorded and transfered to the robot is described in section 8.2. Furthermore, in Appendix Figure C.4 an example is given that shows the difference between a pointing gesture with and without natural movements designed for the iCub. For the eye-neck coordination, an already implemented gazing module is used that is created by the iCub community. It provides human-like eye-neck coordination, and performs saccades using the given 3D coordinates as a parameter. After providing these parameters, the robot tries to gaze at the given point as quickly as possible with the eyes only. After a short delay and at lower speed the head of the iCub follows the movements of the eyes so that in the end the head as well as the eyes are facing the same spot. An additional feature is that this module automatically sets the divergence of the eyes depending on the given 3D point. This behavior adopts human eye-neck coordination making this module perfect as one of the DBCs.

6.3 Idle Movements

Idle movements are all those movements produced when there are no functional movements to be carried out. Humans normally do not stand completely still most of the time. The body is always in motion, even if the movements are not necessary purposeful. There movements range from subtle movements like breathing, to balance shifts from one leg to the other, to facial expressions, to whole body movements like stretching or scratching the head.

According to the theory of form and function in the robot (section 2.2.3) it is possible that missing idle behavior in robots might evoke expectation mismatches when a human-like body is motionless. Furthermore, Song et al. [146] proposed that without motion, users cannot decide if the robot is turned off or damaged. Because of this we decided to look closely at the possible effects associated with idle motions in HRI. The DBCs created include a breathing animation that produces continuous motion in the robot, as well as some more complex idle movements that are produced randomly. The random behaviors include looking around and a stretching. All three behaviors will be explained later on. The principles of natural movements were used as the basis for all movements and more realistic body postures. Beside these motions, a facial expression behavior was integrated to show expressions from sad to happy depending on whether people looked at the robot or not. These DBCs are visual human-like cues that use elements from natural movements DBC. When producing idle movements, it is not enough to produce them with adequate timing. They must also be natural in appearance to fit the overall robot design.

In the area of social robotics, Song et al. [146] observed peoples idle behavior and used it to create idle behavior for their robot. They said that "If a robot does not make any motion in standby state, users may feel that

6. DYNAMIC BACKGROUND CUE PATTERN

the robot is being turned-off or even out of work". This shows that it is advisable to implement idle motions on robots to prevent misunderstandings. The Geminoid HI-1 [137] is another example of a robot that performs idle behavior and looks around after a while. The Geminoid DK can also perform a breathing behavior. In the field of virtual avatars, the topic of idle motions has been discussed for some time. For example, in many recent video games, idle motions are used to make virtual characters appear more vivid. The characters start performing idle movements one their own when the player does not do anything for a while. Egges et al. examined how to improve idle motions for virtual characters. On the one hand, they propose a novel animation approach to create more realistic idle motions that are more flexible and adaptive to previous and subsequent motions by using Principle Component Analysis [47]. On the other hand, they invented a system to generate idle motions while considering the emotional state a robot is in [46]. The advantage of using adaptive and flexible solution is that each animation does not look exactly the same, and there is variation in the motion of the robot. Using idle motions has two goals. First, these can prevent the robot from appearing turned off or damaged. These cues signal the observer that the machine is turned on and that interaction is possible. Second, these DBCs, like all the others, makes the robot appear more lively. As was shown in the first study, breathing can make the robot appear more human-like.

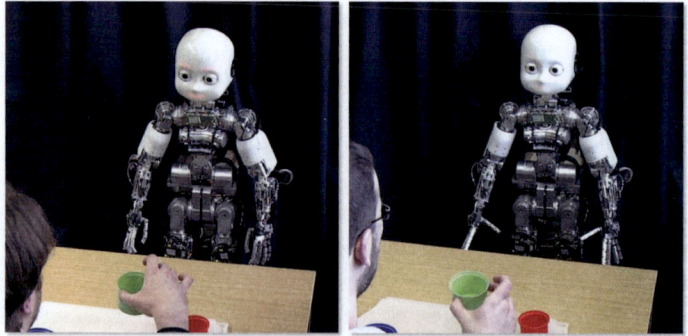

Figure 6.4: Social (left) and non-social (right) posture of iCub

The pattern that is used for these DBCs should generate a continuous motion that prevents the robot from appearing completely still. If there is conflicting functional motion, the idle behavior is turned off so that it does not affect the functional behavior. For example a functional head movement should not affected by the breathing behavior of the upper body, and can be generated in parallel. A reaching motion to grasp something, on the other hand, would suppress the breathing motion.

As a starting point for the breathing and all other idle motions, a natural resting position was created with which all animations start and end. This posture is not as stiff as that found in most default robot resting postures. The arms and legs are bent a little, and the hand posture resembles a human hand at rest. Figure 6.4 shows both social and default postures. Here it can be seen that the right posture is much more stiff, and appears less dynamic.

The breathing animation is a variation of the social resting position that produces slight movements that simulate human breathing. As the iCub is not able to perform real breathing motions, such as lifting the chest, an alternative strategy had to be designed. As a starting point, the social resting position was used from which the upper body was bent towards the front using the hip joint to simulate breathing out. In addition, the arms were rotated at the shoulder joints so that the arms had the same orientation when viewed from the side. The arms were also raised to the sides during the breath. This resulted in a realistic simulation of a breathing motion. All joints only varied a few degrees but this was enough to create a recognizable motion. Figure 6.5 illustrates how this animation looks when viewed from the side. Frequency and the speed of single breaths are parameters that can be modified for this behavior.

While the breathing behavior is produced, idle behaviors are introduced at random intervals to make the idle motion appear more lively. Two different motions have been implemented so far, and the frequency with which these occur can be set by a parameter. The first motion is a "look around" movement. Here the robot looks to one side and then to the other by using the whole upper body, including turning the shoulders and moving the arm in an adequate way. An image of this behavior can be found in Appendix Figure C.2. The other motion is a stretching motion where the robot leans the whole upper body to the left and right, and leans the head in the opposite direction to perform a stretching of the shoulder muscles (See Appendix Figure C.3). For all random animations no key frames are used as described in section 7.10. This can lead to more realistic motions because it is often unnatural if all joint motions end or change direction at

6. DYNAMIC BACKGROUND CUE PATTERN

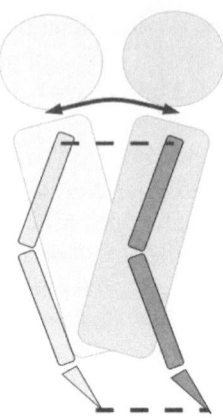

Figure 6.5: Breathing animation for iCub. iCub leans forward and backward by keeping the arm positions in parallel. The movement is exaggerated in this figure for illustrative reasons.

the same time.

To generate facial expressions, a face recognition system was used to detect faces. Each time a human looks in the direction of the robot it becomes happier and it becomes sad if no one is paying it any attention. Depending on the amount of happiness, the robot displays a happy, neutral or sad facial expression.

By using all these DBCs, the goal is to increase the liveliness of the robot by not allowing it to become stiff, and providing varying motions that includes some random behavior.

6.4 Gaze

The gaze of a person is a powerful tool in communication. Much information can be transmitted by the eyes and gaze. The eyes can reveal character traits, elicit joint attention or show emotions. The common saying "The eyes are the window to the soul" supports the idea that the eyes play an important role in HHI, and give clues about internal states. Gaze, as one of the eye's signs, will be examined within this section in more detail. Humans need the eyes to perceive the visual input of their environment. Where human gazes can say much about his internal states. For example, a gaze to a certain object shows interest, or during direct conversation, the gaze can reveal whether the person is lying. The behavior that is focused on here is that produced to reduce discomfort in direct interaction. Humans normally do not stare into the eyes of another person during an interaction. Mutual gaze is interrupted from time to time to fixate on another spot. The duration of mutual gaze varies depending on multiple factors such as the persons character or distance. If one person holds the mutual gaze too long, the other person's anxiety rises and he feels stared at [2]. Currently, robots are able to detect faces and fixate on them to let the human user feel more involved in an interaction. But when following a human face, the robot should break the mutual gaze after a while the way people do. It is hypothesized that a robot that interrupts the direct gaze is perceived as more friendly and familiar. Furthermore, if different patterns of this behavior are used, it may be possible to vary the perceived character traits. The DBC that is used as gazing behavior mainly uses face detection that fixates the face of the human. After a randomly chosen time, the robot terminates the mutual gaze and fixates a salient point in the environment for a short time before gazing back. This DBC is a human-like visual cue mainly used in direct Human-Robot Interaction. If no face is present, the robot randomly gazes at salient points in its environment. To evaluate this DBC in detail an online study was carried out that compares three different variations of this behavior.

Comprehensive research has already been conducted on human gaze. For this chapter, the following findings are relevant. Argyle et al. investigated the duration of mutual gaze depending on distance [3]. Furthermore, they gave a detailed review of mutual gaze duration [2]. When interacting with another person, mutual gaze is not held the entire time. After a while the gaze is broken up and for a short time another point is fixed. This is because a person starts to feel anxious when mutual gaze is too long. When gazed at by another person, there is a certain duration of mutual gaze that is tolerated before growing anxious. If there is no gaze at all, the interaction partner does not feel involved in the interaction. The duration tolerated depends on many different factors. The first factor is individual difference like gender, culture or character. Furthermore, the tolerated duration increases with greater distance, if person A likes B more and if the discussed topic is less intimate. In addition when they are listening, people gaze more at the other person, than when talking themselves. Bee et al. [8] found that virtual avatars are rated as more dominant with an

6. DYNAMIC BACKGROUND CUE PATTERN

upwards oriented head in combination with direct eye gaze. In addition, people that show direct gaze are rated as more dominant than those with averted gaze [100].

In recent years, some researchers began applying human gaze behavior to robots. They evaluated how human gaze within social situations compared to non social situation [169] to adapt this behavior for robots. Others used mutual gaze [24] to achieve the same goal. Poel et al. [122] researched if the gazing behavior of the iCat increased the believability and likability of the robot. To investigate this, they implemented some gazing behavior including blinking, gaze shifts and so forth. Although they did not rate every single cue, they did find that the robot with gazing behavior was seen as more confident and honest even though the behavior was very subtle and was not recognized by the participants. Minato et al. [107][108] investigated the point when humans break eye contact when thinking. They showed that humans behave the same way when interacting with a human-like robot. Garau et al. found out that virtual avatars that use random gaze are perceived as less lively [56]. In addition, they found that if a robot is human-like, a random gaze is less realistic. Although this effect is not significant when using an avatar that is less human-like [55]. Mutlu et al. [112] state that a human can better remember details of a story when gazed at often by a robot that tells the story. Furthermore, Huang et al. [76] invented a Robot Behavior Toolkit that produces humanlike gaze, and found that humans recall information better, and are better at locating objects when the robot uses human-like gaze. In their study both functional and background cues were used for gazing. The robot gazed unintentionally to the desired object which provided task relevant information (functional cue) which may have provided clues in the object finding task and therefore improved performance. All these related work shows that gazing is a relevant social sign for humans as well as for robots.

As the gazing behavior described by Argyle was not investigated separately for social robots, we did so during an online study with iCub. For this, a DBC was created which main intention was to design a gazing pattern that could be used during direct face-to-face interaction where the robot establishes mutual gaze. According to Argyle, it is hypothesized that if a robot held mutual gaze during an entire interaction, it would provoke anxiety. Therefore the DBC is extended to a social gazing an was modified so that it automatically breaks up direct gazing after a while. Furthermore, the robot does not look in random directions even when looking

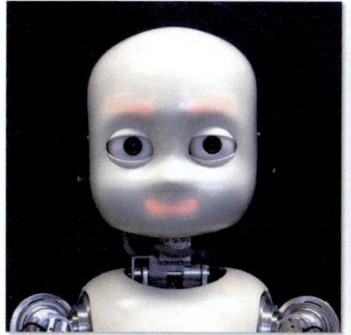 

Figure 6.6: Different stages of iCub performing the gaze DBC)

away. When the robot fixates another point, than a face, a saliency module is used that choses a plausible gazing direction for the observer. This is typically something in motion or colorful objects.

As the proof of concept study showed, there are some problems with the old gazing behavior, such as not establishing mutual gaze or appearing random. The behavior that is introduced here should solve most of these problems, by more involving the human in the interaction and being less random. The way this improved the gaze shall be investigated in the online study. The second problem concerned eye-neck coordination. This new DBC introduced eye-neck coordination to produce a human-like gaze shift which was much more natural. In Figure 6.6, two states of the gaze DBC are shown. On the left side, the iCub fixates on the human user. On the right side, the robot has chosen a salient point in the environment that it is gazed at. The timing when the robot breaks the mutual gaze is randomly chosen depending on the parameter that can be defined for this DBC. The goal of using this DBC is to make the robot more lively and familiar.

Three parameter were used to configure this DBC. On the one hand, the mean duration of the mutual gaze, and the variation can be set as for the blinking behavior. Furthermore, the duration, when the robot again establishes the mutual gaze, can be configured. What parameters are optimal depends on different factors, especially for the duration of mutual gaze. To figure out what parameter were the best for the desired situation and character of the robot, an online study was carried out to investigate the effects of different gazing patterns on the perception of the robot.

6. DYNAMIC BACKGROUND CUE PATTERN

As with blinking, this DBC has the disadvantage that it can influence the vision of the robot. This behavior produces relatively large gaze shifts away from the point of interest. During the gaze shift, a parallel vision system, such as face detection, must address this behavior, or is in the worst case, not functional during the gaze shift.

6.4.1 Online Study

This online study was carried out based on the findings from the proof of concept study. It showed that it was necessary to improve gazing behavior. To achieve this, the social gazing behavior introduced in the last subsection was designed to better involve the user in the interaction. In the last study, many participants complained that the robot did not look into their eyes, and appeared very passive. The new behavior fixates the faces of the human user and follows its movements. To help make the participants feel more comfortable, the robot breaks up mutual gaze as was described by Argyle et al. [2]. The open question is what durations for mutual gaze are adequate within HRI. The question is complicated by how many factors currently remain untested for humanoid robots. In this study, three different durations were tested to measure their effect on HRI during a face-to-face cooperative task while sitting at the same table. By choosing this scenario, variations dependent on distance and situation were adjusted to the setting used for this thesis. For different situations or distances other parameter could be more optimal.

First, we tested which parameters were capable of making the robot more friendly and familiar (Hypothesis 1). We hypothesized that different parameters could create different character traits for a robot (Hypothesis 2). The second goal was to test whether these modifications could be productive, and to find the parameters that best fit the desired robot personality. We tested three durations for mutual gaze: infinite, medium and exaggerated.

6.4.1.1 Setup

This study was designed as an online study that could be completed on a personal computer using any standard browser. A PHP-Script was written that was able to show videos and questionnaires whose results were saved in a mySQL-Database for further evaluation. During the study, the participants had to watch videos of the iCub, and them using 10 items. A within-subject study design was chosen such that all participants viewed all three conditions in random order. The webpage that displayed the videos is shown in the Appendix Figure A.5. After each video, the participants had to complete a 10 items survey, resulting in total 30 items. After viewing the videos, participants completed a general questionnaire containing personal data. The link to the study was distributed by E-Mail and social networks to reach a broad variety of people. On the first page of the study, the participants were instructed to watch the three videos twice carefully and to make sure that sound was on. In this study, no sound was used but the instructions need to be consistent for the subsequent studies. Furthermore, participants were told to rate the videos based on their gut feelings, and that they should do the study only when they were alone to avoid disturbances. The original instructions in German can be found in the Appendix Figure A.4

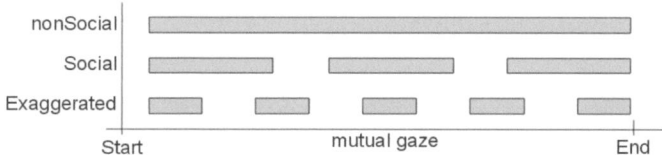

Figure 6.7: Timing of mutual gaze during the online study

There were three videos of the robot that should be rated. Each video showed a different gazing behavior, with the iCub visible down to the shoulders(see Figure 6.6). The angle of the camera that recorded the iCub was chosen because it matches the distance and angle from which users would see the robot in a face-to-face cooperative task. The videos started with the iCub looking into the lower left corner of the screen. During the first second, the robot established mutual gaze and fixated the human (Figure 6.6(left)). After a given time span, the robot broke eye contact and looked elsewhere (Figure 6.6(right)). Thereby the gazing direction varied from time to time. The three conditions were the following. First condition was the *infinite* condition. Here mutual gaze lasted an infinite period of time, meaning that the robot did not break eye contact and stared at the participant during the whole interaction. This condition represents the gazing behavior that is often used for robots because they tend to not use

6. DYNAMIC BACKGROUND CUE PATTERN

a gazing strategy. The second condition was the *medium* condition which was oriented at the parameters that are typical for humans in the same situation. The typical duration was 3-10 seconds when listening. For this condition, four seconds were chosen. The last was the *exaggerated* condition and represents the other extreme by breaking eye contact very often. This condition was designed to have durations shorter then the minimum 3 seconds durations normally used in HHI. Here ca. 2 seconds were used. Figure 6.7 illustrates all three conditions. The blue boxes represent mutual gaze, whereas the spaces between stand for the time the robot fixates another point within the environment.

6.4.1.2 Measurements

The measures for this study were questionnaires with 10 items that could be divided into three different groups. Responses were forced-choice using a 6 point scale. Furthermore, each side was named by opposing extrema of each dimension (e.g.:too fast/too slow). The latter part always had the highest value (too fast = 1; too slow = 6). The study was designed such that the questionnaires could only be seen and completed after leaving the page that contained the video. This ensured that the participants could not watch the video again after seeing the questions and vice versa.

The first group of items concerned the character traits of the robot and consisted of five items. First, the three-factor emotion model (VAD-model) designed by Russel & Mehrabian [133] was used. It consists of three dimensions: valence, arousal and dominance. This model provided three items (positive/negative, calm/aroused and dominant/submissive) for the study. In addition two items from Eyssel et al. [52] were also used (nervous/relaxed and attentive/disinterested). The second group consisted of one item (not human-like/very human-like) and provided the option to comment on the choice using simple language. The last group consisted of four items that considered the functional appearance of this DBC (too fast/too slow, pleasant/unpleasant, disturbing/not disturbing and the mutual gaze was... too long/too short). All the original German items can be found in the Appendix Figure A.6.

After filling out the questionnaires for all three conditions, participants also completed a personal data form. This data consisted of age, gender, native language, job, highest degree earned, computer experience, robot experience and if they had already participated in a study with iCub. In addition the Big5 scale was used.

6.4.1.3 Results

The results of this online study are the following. 29 participants (16 female, 12 male, 1 not specified) took part in this online study with age ranging from 16 to 63 ($M = 33.62$, $SD = 12.517$). The participants rated their own computer and robot experience on a scale ranging from 1 (none) to 5 (very much). They had a mean computer experience of 3.58 ($SD = 1.065$) and a mean robot experience of 1.86 ($SD = 0.915$). Because of the study design all participants rated all three conditions. As the study was carried out in Germany all participants were German expect one.

To evaluate the online questionnaires Friedmann Tests were carried out to test for significance. In the case of significance, a Wilcoxon Signed Rank Test was carried out as a Post-hoc test to see which conditions differed significantly. When using the Wilcoxon Test the significance level decreased to $sig\_level/number\_of\_tests = 0.05/3 = 0.0167$. Within the first group, the character traits showed significant results. There were no significant results for the Friedmann Test for the items **attentive/disattentive** ($\chi^2(2) = 1.407, p = .495$) and **positive/negative** ($\chi^2(2) = 2.96, p = .228$). For the item **nervous/relaxed** there were two significant differences. The infinite ($Mdn = 5$) as well as the medium ($Mdn = 5$) condition differed significantly from the exaggerated condition ($Mdn = 2$). The exaggerated was rated as more nervous than the infinite ($Z = 3.6, p =< .001, r = 0.69$) and the medium ($Z = 3.68, p =< .001, r = 0.69$). There was no difference between the infinite and the medium conditions, $Z = 0.4, p = .687, r = 0.07$. The item **calm/aroused** reveals similar results. The infinite ($Mdn = 2$) and the medium ($Mdn = 2$) condition showed significant differences from the exaggerated ($Mdn = 5$) condition. The exaggerated gaze movements were rated as more aroused than the infinite ($Z = 4.17, p =< .001, r = 0.78$) and the medium ($Z = 3.92, p =< .001, r = 0.74$). The infinite and the medium condition showed no significant differences $Z = 1.2, p = .231, r = 0.22$. The last item in this group, the **dominance/submissive** item, showed significant differences in the infinite ($Mdn = 3$) and medium ($Mdn = 5$) conditions, $Z = 2.83, p = .005, r = 0.53$ there are no further differences between them and the exaggerated condition ($Mdn = 4$).

Within the second group the item **not human-like/very human-like** showed one significant difference between infinite ($Mdn = 4$) and medium

6. DYNAMIC BACKGROUND CUE PATTERN

($Mdn = 5$), where the medium condition was rated as more human-like $Z = 2.7, p = .007, r = 0.5$. There were no differences between infinite or medium and the exaggerated condition ($Mdn = 4$).

In the group of the functional items, the participants rated **too long/too short** as significantly different among all three conditions. Here the mutual gaze in the infinite condition ($Mdn = 1$) was rated as too long compared to the medium condition ($Mdn = 4$), $Z = 4.21, p =< .001, r = 0.8$, as well as to the exaggerated condition ($Mdn = 4$), $Z = 4.51, p =< .001, r = 0.84$. Furthermore the gaze of the exaggerated condition was rated as too short compared to the medium condition, $Z = 2., 42p =< .001, r = 0.46$. The item **too fast/too slow** reveals two significances. The infinite ($Mdn = 4$) as well as the medium ($Mdn = 4$) condition differed significantly from the exaggerated condition ($Mdn = 2$). The exaggerated condition was rated as too fast compared to the infinite ($Z = 3.99, p =< .001, r = 0.77$) and the medium condition($Z = 3.8, p =< .001, r = 0.73$). There is no difference between infinite and medium condition, $Z = 1.6, p =< .001, r = 0.31$. For the **pleasant/unpleasant** item the medium condition ($Mdn = 2$) was rated as more pleasant than the infinite ($Mdn = 4$), $Z = 3.31, p = .001, r = 0.63$, and the exaggerated condition ($Mdn = 4$), $Z = 3.48, p =< .001, r = 0.65$. There was no difference between the infinite and exaggerated conditions, $Z = 0.37, p = .711, r = 0.07$. The last item for this group, the **disturbing/not disturbing** item, showed two significant differences. The participants rated the medium condition ($Mdn = 5$) as less disturbing than the infinite ($Mdn = 4$), $Z = 2.48, p = .013, r = 0.49$ and the exaggerated condition ($Mdn = 4$), $Z = 2.88, p = .004, r = 0.55$. This item was rated without differences between the infinite and the exaggerated condition, $Z = 0.32, p = .752, r = 0.06$. A complete list of all items can be found in Appendix B.2.

In addition to the differences between the two conditions, correlations were evaluated to test if there were coherences between different answers in one condition. To compute the correlations, Spearman's Rank Order Correlation was used. Here only the most important correlations are presented. In the infinite condition the robot was rated as more **disturbing** when the participants were older, $r_s(29) = -.37, p = .048$. Furthermore, there was a negative correlation between **computer experience** and **arousal** in the medium condition. The robot was rated as more **calm** when the participant has more **computer experience**, $r_s(29) = -.509, p = .005$. In the area of robot experience there was a positive correlation for **re-**

laxed in the medium condition, $r_s(29) = .454, p = .013$. Furthermore in the exaggerated condition there were correlations between robot experience and **"very human-like"**, $r_s(29) = -.585, p = .001$ as well as **"less disturbing"**, $r_s(29) = .368, p = .050$. Other interesting correlations appeared for those participants that rated themselves as **convivial**. In the infinite condition those participants tended to rate the gaze of the robot as less **unpleasant**, $r_s(29) = -.444, p = .016$. In the exaggerated condition more convivial participants rated the robot as more **dominant**, $r_s(29) = -.369, p = .049$.

Within the questionnaire, the participants were asked to reason their decision on the item not human-like/very human-like. In the infinite condition, both given comments were negative. One said that the robot was staring and the other described the behavior of the robot as provocative. In the medium condition, three of the four comments were positive. Those participants rated the robot as human-like because of its mutual gaze, changing gaze direction and the existence of nonverbal signals. The fourth comment described the robot as shy because of its gaze was mainly directed at the floor. In the exaggerated condition, six comments were given. One positive comment says that the behavior of the robot fit to the expectations. A neutral comment stated that it depends on the situation how human-like the robot should be rated. This participant suggested that the exaggerated gaze could fit to an excited discussion. The remaining comments were mainly negative. Here the participants stated that the movement of the robot was too fast and some said that the eye-neck coordination was not adequate in this condition. One participant describes the robot as inattentive and jittery.

6.4.1.4 Discussion

In this online study we investigated how the gaze DBC influenced how the human participant perceived the robot, and if this DBC made the robot appear more familiar and human-like. The first two Hypotheses presented in section 3.3 are tested.

One part of the second Hypothesis could be confirmed. Single cues were capable of creating character traits depending on their parameter. In the infinite condition, the robot was rated as more dominant than in the medium condition, and as more relaxed and calm than in the exaggerated

6. DYNAMIC BACKGROUND CUE PATTERN

condition. This shows that participants associate staring behavior with dominance as in HHI. In addition, the low amount of movements allowed the robot to appear relaxed and calm. When looking at the absolute ratings, the character profile of this behavior was relaxed, calm and dominant. This was also supported by the qualitative data. Both comments on this behavior stated that the robot was provocative, and that it stared all the time, which again showed the perception of dominance in the character of this robot. In the exaggerated condition, the opposite effects can be found. Due to the large number of gaze shifts, the robot appeared more aroused and nervous then in the other two conditions. Indeed the robot was rated as less dominant because it broke up mutual gaze although this behavior was rated by some participants as inattentive and jittery. As expected, the ratings in the medium condition stood between the results of the other two conditions. This behavior was rated as more relaxed and calm than the exaggerated condition and less dominant than the infinite condition. Referring to the absolute values of the data striking character traits for this behavior were relaxed, calm and submissive. In addition, one comment in the qualitative data rated this behavior as shy, which supports the submissive appearance of the robot. In all three conditions, there were no differences for the items attentive/disinterested and positive/negative. In all cases, both items were rated as neutral without any tendencies to one or the other extreme. As the second hypothesis was supported, the VAD dimensions could be used to sort the behavior of the robot into the three-factor emotion model.

The goal was to create a friendly robot for a cooperative task. For this, the medium condition seems to fit best. The exaggerated behavior of the robot appeared too nervous which could also arouse the participant too much. The infinite behavior could make the human feel inferior, since the robot stares all the time. The medium behavior has all the character traits that support the robot as a friendly cooperative robot. In other scenarios, other behavior could be more effective. For example, when using the robot as a coach for sports, it might be better to let the robot appear more dominant.

The first Hypothesis could be supported as well. The gaze DBC increased familiarity and the perceived liveliness of the robot when using the right parameter. The medium condition received the highest rating for human-likeness, and differed significantly from the infinite condition. For the functional items, the medium condition was best rated as well. The infinite condition was rated as having too long mutual gaze and too slow movements. The exaggerated condition as having too short mutual gaze and too rapid movements. The medium condition was rated between the other two conditions. In terms of the absolute values, they show that the medium condition was a little too slow and mutual gaze a little too short but nevertheless, it was better rated than the other two. In another iteration, the timing of the gazing can be improved further. In terms of pleasantness and disturbing behavior, the medium condition outperformed the other conditions significantly in being more pleasant and less disturbing. The qualitative results support these findings. In general, the medium condition received the most positive results compared to the others. Interestingly, in the exaggerated condition the eye-neck coordination was criticized, whereas there was no criticism for the medium condition even though the same movements were used for gaze shift. Possibly other parameters for eye-neck coordination were expected when the robot performed a more nervous gaze behavior. These results showed that, except for small timing issues, the gazing strategy used in the medium condition clearly improves the gazing strategy used in the proof of concept study. In terms of the gazing behavior for HHI described by Argyle it can be said that in HRI the same effects take place. Mutual gaze that lasts too long is perceived as unpleasant as well as mutual gaze that is too short. First, because of feeling of inferiority (high dominance value), and later because the robot appears nervous and aroused.

In addition some interesting correlations occur in this online study. First there is a correlation between age and the perception of the robot. During the infinite condition the staring behavior was rated as more disturbing when the participants were older. But as there was no correlation in the other two conditions, this effect should have no negative influence when using the medium parameter in further research. Another negative correlation occurred between computer experience, and the item calm/arousal. Participants with more computer knowledge rated the robot as less aroused in the medium condition. Participants with greater robot knowledge rate the robot as more relaxed in the medium condition and as less human-like in the exaggerated condition. This shows that experience with technical devices can influence how social robots are perceived and that this should be considered when designing them. In this case, the effects were not disturbing and even supportive when using the medium parameter. In addition, people that rated themselves as more convivial rated the robot as more pleasant in the infinite condition, and as more dominant in the exaggerated condition. This is against the general trend, and further supports the decision to choose the medium condition. Further, it shows that the individual differences in the characters of the users should not be ignored. These correlations are only some examples. There are even more correlations in the data which refer to similar effects.

6.5 Lip-Movement

The lips of social robots are becoming more and more realistic, as for example those of Geminoid (see Figure 4.4) or Flobi (see Figure 4.5). In addition, more iconic lip solutions can be found in the iCub, for example (see Figure 6.8). But the existence of lips may evoke certain expectations of robots, such as the use of lips during speech. This is why the lip DBC was created to test if the use of lips during speech could influence the user, and improve the interaction. This DBC aimed to create a lip synchronous speech synthesis. As soon as the robot spoke, a corresponding lip movement was created that used lip postures (visemes) like those of humans when speaking. For example, when uttering an "a" the mouth was open and when uttering an "m" the lips were closed. This DBC is a visual human-like cue. It is not auditory because speech is not part of the DBC, because it is mainly a functional signal

Lip movements itself are not necessary to understand another person. For example, when talking via telephone there are no additional signs except for the auditive one. But lip movements can help to get additional information. Through lip reading it is possible to understand the speaker better in noisy environments, or it is possible to identify who is speaking among a group of people. For robots, the problem is that it is not really speaking. In some cases, the speech that is synthesized is produced by a loudspeaker that is not located in the head of the robot, where sound normally comes from with people. This makes it harder to accept the robot as the producer of the sound. Lip movements could help in such a situation. For example, humans accept the puppet of a ventriloquist as the speaker although the person produces the speech. The puppet only moves the lips synchronously to the speech.

Production of speech-synchronous lip movements has already been implemented and evaluated for its recognizability. For virtual avatars Waters et al. [156] introduced an algorithm to automatically generate fitting lip movements for virtual characters out of plain ASCII text. To achieve this,

they generated a phonetic transcription of the text that was used to get the synthesized audio stream. In parallel, this transcription was used to generate a viseme sequence that was displayed by a virtual character. Taking a different approach, Bozkurt et al. [13] evaluated methods to better generate synchronous lip movements out of pre-recorded speech by using HMM based recognition systems. In the field of robotics, one good example of lip synchronous speech synthesis is given by Breazeal. For Kismet [21] a system was implemented to produce realtime lip movements using DECtalk [157]. In addition, she extended the speech system of Kismet [16] to produce emotional speech and evaluated how well people understand the emotional cues. Burkhardt [28] made the same evaluation for pure audio without visual signs. Both reached adequate recognition results but observed some problems for very similar emotions like fear and surprise
Another way of using lip movements for robot or computer systems is automatic lip reading. Bruce [25] gave a good review of this topic, and described how it can be used for robots. He argued that lips provided a lot more information than speech alone. Considering lip movements can help to understand what is said. Visser et al. [151] introduced a system that used visual information for automatic lip reading which could be used to improve audio based speech recognition.
This shows that the lips provide a great deal of information that can be perceived by a human. To prevent an expectation mismatch, it is advisable to implement lip movements. Furthermore, these have some obvious advantages such as speaker identification. Up to now speech-synchronous lip movements have only been evaluated with respect to recognizability but not to how they affect the perception of the robot in terms of social features. To fill this gap, we carried out an online study.

In the proof of concept study, some participants criticized that the robot produced no speech, We therefore included and improved this feature in the presented system. Based on the advantages that are described above, speech-synchronous lip movement was created as a DBC to fulfill the expectations of the user. On a technical level, this DBC orients at the implementation of Waters et al. (see above). Here the MARY Text-To-Speech System (version: 3.6.0) is used [140] to generate the audio output based on a plain ASCII text. MARY is able to export a phonetic transcription from text. Phonemes are the smallest units that are used in speech. For example an "a" is one phonetic unit that is used for example in "start". This phonetic transcription has an XML-Structure and uses SAMPA variation

6. DYNAMIC BACKGROUND CUE PATTERN

Figure 6.8: Possible visemes of iCub (`close` and `open`)

to represent each phoneme. SAMPA [162] is a computer readable phonetic alphabet that consists only of ASCII characters to make it possible to use on a computer. Older phonetic alphabets like IPA use characters that are not present in ASCII code. For example the "a" in "start" is presented as an "A" in SAMPA. By using this phonetic transcription, it is possible to extract the length and timing of each phoneme within the synthesized text. Out of this, a corresponding stream of visemes is created. A viseme is the corresponding lip posture that is used when uttering one phoneme. There are always several phonemes that have the same viseme. For example "b" and "p" use the same viseme. With this stream of visemes, it is possible to perform the right lip movements that fit the speech produced by the robot. Figure 6.8 shows the two visemes the iCub is able to display.

One open question is how to convert the phonemes into the corresponding visemes, and how to display these visemes in the robot. There are several studies that consider speech and viseme mapping. For example Aschenberner et al. [4] presented a viseme mapping for German language using BOSS as the phonetic alphabet. Furthermore, Waters et al. [156] used DECtalk for their phoneme and viseme mapping. They also presented a set of images that showed the visemes for virtual characters. For this research, the work of Aschenberner et al. was used as the foundation because the first speech module was only able to speak German. In a second iteration, the English language was included. The modules used to realize the speech synthesis were in both cases Hidden Marcov Model based. For the English speech we used model "english-hmm-bdl v.3.5.0 (TractScaler:1.1; Volume:3.0; F0Scale:2.0; F0Add:100.0; Rate:1.1)" and for German speech we used "german-hmm-bits3 v.3.5.0 (TractScaler:1.2; Volume:3.0; F0Scale:2.0; F0Add:60.0; Rate:0.9)". All parameter for both models were chosen so that the generated speech fit the appearance of the robot. This means that a friendly and child-like voice was used because the age of the robot was rated as very young in the proof of concept study. The complete mapping can be found in the Appendix B.3. The viseme labels are taken from Aschenberner et al. [4] which are presented on Figure C.1 in the Appendix.

The lip constructions of robots often have limitations compared with human lips. Therefore it was necessary to find a way to display the visemes in an understandable way because they could not duplicate human lip postures. The first tests for the lip movements were done on the Flobi robot (see section 4.5). This robot has mechanical lips that are attached via magnets to the actuators that are placed beneath the plastic cover. The first tests showed the high demands on the hardware during speech. The frequency with which the lips have to be moved is about 10 visemes/sec during normal speech. In the worse case, the robot has to open and close the mouth 10 times a second, but it can be less if two phonemes are similar and there is no movement needed. Nevertheless, the load for the hardware is very high. When using the Flobi, overheating problems occurred often caused by high friction at the magnetic connections during continuous movement. Furthermore, the noises that were produced by the motors when moving at the desired speed were very loud, and disruptive when listening to the speech the robot produced. The iCub has fewer degrees of freedom in the lips but has no problem producing the high frequency of phonemes because it uses LEDs instead of mechanical lips. This robot can only display two visemes (see Figure 6.8). An open mouth and a closed one. When the iCub receives the command to present one of the desired visemes, an additional mapping realizes the production of fitting lip postures that correspond to the given viseme.

Unfortunately, it is not possible to display visemes between open and closed lips with the iCub. As such, the robot is not always able to display the correct visemes for the corresponding phonemes. When comparing with movies and animations it seems that it is not necessary to exactly match the visemes. Many movies are synchronized into other languages so that the actors move their lips to English, for example, only to hear German

6. DYNAMIC BACKGROUND CUE PATTERN

being spoken. Most people do not realize that the lips and the speech do not fit when there are no huge deviations. Furthermore, in animated movies, the characters often use fewer different visemes than are necessary for human speech. The mapping from visemes to lip postures can be found in the Appendix B.3

The lip DBC itself has no parameters, because the lip movements are extracted from the text that will be uttered. For testing purposes, an alternative lip DBC has been created that randomly produces lip movements based on pre-defined average visemes/sec value. This makes it possible to test the effects of different lip movements on the perceived familiarity and human-likeness of the robot. To evaluate this effect, another online study was carried out that is presented in the following section.

Up to now, there was no emotional speech implemented in the system, but as an emotion system for facial expressions already exists, the goal is to implement emotional speech in the future.

6.5.1 Online Study

In the previous section, speech-synchronous lip movement was presented. This behavior is a simple adoption from human lip movement behavior. In this section, an online study is presented that evaluates how these lip movements influence how the robot is perceived, and if the adoption of the human behavior is adequate for HRI in a cooperative task. The structure of the online study that is carried out to test the different lip movements is the same as the study used to evaluate gaze movements. Only the videos and some items were changed. The reason for evaluating this DBC is that in the proof of concept study it was criticized that the robot did not give any feedback. Adding speech as new feedback to the system, is one component to fulfilling this expectation. When adding speech it may be advisable to add lip movements which may cause expectation mismatches if they are missing. We will test if it is necessary, and what the differences between different behaviors might be.

The goal of this study is to test if the DBC is capable of making the robot appear more friendly and familiar (Hypothesis 1). In addition, we tested if it is possible to change the perceived character traits of the robot depending on different parameters (Hypothesis 2). In addition, it should be possible to choose parameters that fit the desired character for the robot. For this study three conditions have been used: noLips, synced, exaggerated.

6.5.1.1 Setup

The setup used was the same within-subject study design as for the gazing study presented in section 6.4.1. This study was carried out as an online study that each participant could do on their own using a browser. The study began with short instructions that can be found in the Appendix Figure A.4. After reading the instructions, three videos were rated (See Figure A.5) with a questionnaire that was displayed after watching each video. Each video belonged to one condition, and was presented in a random order. The questionnaire contained 10 items as described in the next section. After filling out all 30 questions, an additional questionnaire with personal data was filled out. The only difference to the gazing study was that different videos were used to evaluate the three conditions for the lip behavior, and that the 10 items for each condition changed slightly. Furthermore, this study required sound to hear the speech of the robot.

Three conditions were created that used different lip movements. In all conditions, the same sentence and timing was used to keep the functional behavior the same. Furthermore, the same camera angle used in the gazing study, was chosen to fit the scenario of a cooperative interaction. The head of the iCub did not move during the video, except for the lips during speech. The sentence that was spoken in all three conditions was "Can you show me the way to the train station? ". This sentence was chosen because it is very functional without carrying any emotional information that could be transmitted by the meaning of the words. In the *noLips* condition, the robot did not move the lips while the speech was played. This condition represented the behavior when the lip DBC is not turned on. In the *synced* condition, the lips were moved by using the synchronized lip movement presented above. This behavior best fit with human behavior. Around about 10 visemes/sec were performed. In the *exaggerated* condition, the alternative lip movement was used to produced random lip movements. Here a frequency of about 50 visemes/sec is used, which is much higher than in the synced condition.

6.5.1.2 Measurements

The measurements for this study were questionnaires with 10 items that could be divided into three different groups. Those were forced-choice using

6. DYNAMIC BACKGROUND CUE PATTERN

a 6 point scale. Furthermore, each side was named by opposing extrema of each dimension (e.g.: too fast/too slow). The latter part always had the highest value (too fast = 1; too slow = 6). The first two groups of these items, as well as the questionnaire that asked for personal data were the same as in the gaze study (See section 6.4.1.2 for further details).

The third group of items consisted of four items that considered functional appearance. These items aimed to investigate if there were obvious differences in the behavior of the robot that were perceivable by the user. For the lip behavior the questions were "The lip movements of iCub were...": synchronous/asynchronous, too fast/too slow, pleasant/unpleasant and disturbing/not disturbing. The original German questionnaire can be found in the Appendix Figure A.7.

6.5.1.3 Results

In this section we present the results of the online study. 33 Participants took part in this study (15 female, 18 male). The age of the participants ranged from 16 to 65 years ($M = 32.73$, $SD = 12.996$). The participants rated their own computer experience on a scale from 1 (none) to 5 (very much) with a mean value of 4.07 ($SD = 0.907$) and their robot experience on the same scale with a mean value of 2.3 ($SD = 1.104$). All participants that attended this study were German speakers.

To evaluate the online questionnaires Friedmann Tests were used to test for significance. In case of significance, a Wilcoxon Signed Rank Test was carried out as post-hoc test to verify which conditions differed significantly. When using the Wilcoxon Test, the significance level decreased to $sig\_level/number\_of\_tests = 0.05/3 = 0.0167$. Within the first group three items had significant results. The items **nervous/ relaxed** ($\chi^2(2) = 5.828, p = .054$) and **calm/aroused** ($\chi^2(2) = 1.820, p = .403$) did not provide significant results in the Friedmann Test. The item **attentive/disinterested** shows one significant result between the noLips ($Mdn = 4$) and the synced ($Mdn = 3$) condition, $Z = 3, p = .003, r = 0.52$. There were no further significant effects between the exaggerated ($Mdn = 3$) condition and the other two conditions. The item **positive/negative** showed one significant result, as well. For this item the noLips ($Mdn = 3$) was rated as more negative than the synced ($Mdn = 3$) condition, $Z = 3.5, p = < .001, r = 0.61$. There was no further difference between these two conditions and the exaggerated condition ($Mdn = 3$). There was another significant effect for the last item of this group. The item **dominance/submissive** was rated as more dominant in the exaggerated ($Mdn = 4$) than in the noLips ($Mdn = 4$) condition, $Z = 2.45, p = .014, r = 0.43$. The one item in the second group had two significant results. For the item **not human-like/very human-like** the noLips condition ($Mdn = 1$) was seen as less human-like when compared to the synced ($Mdn = 3$), $Z = 4.6, p =< .001, r = 0.83$ as well as to the exaggerated ($Mdn = 3$) condition, $Z = 3.78, p =< .001, r = 0.68$. There was no difference between synced and exaggerated, $Z = 2.3, p = .021, r = 0.41$.

In the last group, all items showed significant results in the Friedmann Test. The item **synchronous/asynchronous** showed significant differences between the noLips ($Mdn = 6$) and the synchronous ($Mdn = 2$) as well as between the noLips and the exaggerated ($Mdn = 3$) condition. The synchronous was rated as more synchronous compared to the noLips condition, $Z = 4.32, p =< .001, r = 0.85$. In addition the exaggerated condition was rated as more synchronous than the noLips condition ($Z = 3.74, p =< .001, r = 0.73$). There was no significant difference between synced and exaggerated, $Z = 1.95, p = .051, r = 0.35$. The item **too fast/too slow** was rated as too slow in the noLips ($Mdn = 6$) condition compared to the synced ($Mdn = 3$) condition, $Z = 3.86, p =< .001, r = 0.74$. Furthermore the noLips condition was rated as too slow compared to the exaggerated ($Mdn = 3$) condition, $Z = 4.14, p =< .001, r = 0.8$. There was no significant difference between the synced and the exaggerated condition. The item **pleasant/unpleasant** revealed significant differences between all three condition. The synced condition ($Mdn = 3$) was rated as more pleasant than the noLips ($Mdn = 5$) condition, $Z = 4.31, p =< .001, r = 0.83$ as well as the exaggerated ($Mdn = 3$) condition, $Z = 3.47, p = .001, r = 0.67$. The exaggerated condition was rated as more pleasant than the noLips condition, $Z = 3.05, p = .002, r = 0.55$. For the item **disturbing/not disturbing**, there were significant differences between noLips ($Mdn = 3$) and synced ($Mdn = 5$) and between synced and exaggerated ($Mdn = 5$). The synced condition was rated as less disturbing than the noLips ($Z = 3.52, p =< .001, r = 0.64$) and the exaggerated ($Z = 2.49, p = .013, r = 0.44$) condition. A complete list of all items can be found in Appendix B.4

In addition to the differences between the two conditions, correlations were evaluated to test if there were coherences between different answers in one condition. To compute the correlations, Spearman's Rank Order

6. DYNAMIC BACKGROUND CUE PATTERN

Correlation was used. Here, only the most important correlations are presented. Three interesting correlations occurred for the computer experience. The robot was perceived as less **dominant** when the participants had more **computer experience** in the noLips condition $r_s(33) = .47, p = .006$ as well as in the exaggerated condition, $r_s(33) = .396, p = .022$. Furthermore, in the synced condition, the robot was seen as less **pleasant** with increasing **computer experience**, $r_s(33) = .367, p = .035$. Other correlations occurred for the question of whether the participants rated themselves as more **reserved**. More reserved participants rated the robot as more **human-like** in the noLips condition ($r_s(33) = .434, p = .012$), less **disturbing** in the exaggerated condition ($r_s(33) = .462, p = .007$) and more **attentive** in the synced condition ($r_s(33) = -.348, p = .047$).

Within the questionnaire, the participants were asked to explain their decision for the item not human-like/very human-like. For all conditions, there were comments that the missing head movements were disturbing. In the noLips condition, some participants said that the robot appeared artificial and that the non-existent gaze was disturbing. For the synced condition, one participant said that the voice of the robot was more pleasant even though the same voice was used as in the other two conditions. For the exaggerated condition, there were no additional comments.

6.5.1.4 Discussion

This online study aimed to investigate the effect of different lip movements on how people perceived the robots. We sought to evaluate if these behaviors were able to make the robot more familiar and human-like, and if these behaviors could alter the perceived character traits of the robot. As a basis for this study, the first and the second Hypotheses, introduced in section 3.3, were used. These results were used to improve the lip DBC that was introduced in this section.

The data that were gathered during this study could support the second hypothesis for this DBC. Different lip behaviors seem to be able to create different character traits for a robot. One interesting outcome was that there was no significant differences between the three conditions for nervous/relaxed and calm/aroused. This showed that the lip behavior seemed to have only little influence on how the activity of the robot was rated. In general, the robot was rated as relatively calm and relaxed. A reason for this might be that the robot performed almost no movements in this study. The comments in the qualitative date further support this. The participants stated that they missed the movements of the head in all three condition.

Furthermore, the synced condition was rated as more attentive than the noLips condition. A reason for this might be that the participants accepted the robot more as the speaker when the robot moved its lips. When the robot did not show any connection to the speech, it was perhaps not seen as the speaker, and therefore seen as more inattentive. There was no significant difference in the exaggerated condition for this item but the mean value showed that it was rated between the other two conditions. This means that it was rated as slightly more attentive than in the noLips condition. Whatever might be the exact reason for this effect, it shows that, when using speech with proper lip movements, it increases how attentive the robot is perceived to be.

For the item that refers to the VAD dimensions, two of three showed significant differences: positive/negative and dominant/submissive. The exaggerated condition was rated as more dominant than the noLips condition, and the synced condition was rated as more positive than the noLips condition. The synced condition was chosen to create a friendly, cooperative robot. For the same reasons as for the gazing behavior, the condition was chosen where the root appearance is less dominant and most relaxed. As there was no difference in activity, all conditions are adequate for this character trait. For the dominant character trait, the exaggerated condition had disadvantages. Another reason for this decision is the more positive rating of the synced condition.

Within this study, the first hypothesis could again be supported. The lip movements were able to increase human-likeness and familiarity. The human-likeness was significantly higher in the synced and in the exaggerated condition compared with the noLips condition. This shows that the use of lip movement is an improvement to no movements. The absolute values of human-likeness show that the rating in general is relatively low. One reason for this could be that the robot did not perform head movements, and therefore was rated similarly to the infinite gazing condition in the previous study. In this study, the robot performed the same staring gazing behavior, and therefore the rating was not surprising. Furthermore, the absence of movements was also described in the qualitative data as a problem. In addition, in the noLips condition the staring behavior of the robot was mentioned as disturbing but not in the other conditions where

6. DYNAMIC BACKGROUND CUE PATTERN

the same gaze behavior was used. It seems that the participants focused more on the lips, and therefore did not pay too much attention to gaze. Nevertheless, lip movements cause an effect, and would likely perform well when used in combination with proper gaze.

Within the third group of questions, in most cases, the synced and the exaggerated condition, outperformed the noLip condition. The noLips condition was significantly seen as more asynchronous, too slow, more unpleasant and more disturbing compared to the other conditions. For the absolute values, the other two conditions were seen as very synchronous, not too fast and not too slow, having a medium pleasant score, and are not very disturbing. In direct comparison, the synced condition was rated a bit better then the exaggerated condition. The data show that the synced condition is significantly more pleasant and less disturbing than the exaggerated condition.

This indicates that the synced as well as the exaggerated condition are useful when improving human-likeness and familiarity compared to no behavior. The synced behavior can be preferred when designing a friendly robot as it is desired within this research because of its advantages in pleasantness and lower disturbance.

The comparison between the items when rating the robot, and the personal data revealed additional correlations. Two attributes for the human user seemed to have noticeable influence on how the robot was perceived. First, technical experience correlated with some ratings as it did in the gazing study. In both extreme conditions (noLips and exaggerated), participants with higher computer experience rated the robot as less dominant. Furthermore, in the noLips condition, those participants tended to rate the robot as more relaxed. Surprisingly, participants with more computer knowledge rated the robot in the synced condition as less pleasant. A possible interpretation of this might be that users with greater computer experience are more tolerant to design mistakes because of their experience with technical systems. The experience with technical systems may also cause the social behavior in the synced condition to be seen as more unpleasant because it is not typical for a technical device to behave socially. Similar effects seem to exist for people that rate themselves as more diffident and reserved. Those people tended to rate the robot as more humanlike in the noLips condition and as less disturbing in the exaggerated condition. In addition, more reserved users rated the synced condition as more attentive. One explanation for this could be that more reserved people forgive more mistakes, or non-human-like behavior in an anthropomorphic robot than more extroverted users.

Chapter 7
System Description

In the previous chapters mainly theoretical issues were discussed and user studies were described that evaluated the effects of Dynamic Background Cues on how robots were perceived. In this chapter we describe the underlying robotic system that allows us to generate the DBCs. A system needs to be created that can encompass or wrap the robot in an architecture that makes the robot social by producing intrinsic behavior and movements. Furthermore, an interface should be provided that supports, beside the basic robot interface, high-level functions that can be used by external functional systems. The architecture of this system aims to achieve this by using a Social Layer Architecture (SLA) that encapsulates the robot's hardware and the functional system(See Figure 7.1).

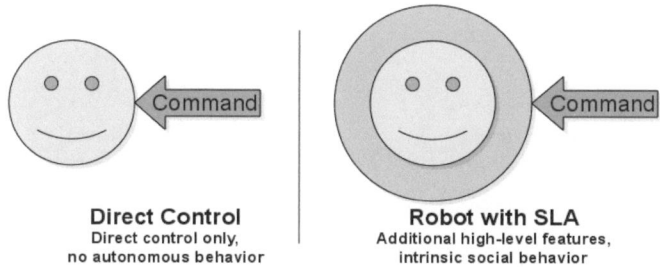

Figure 7.1: Different strategies to communicate with a robot

The two developer centered criteria presented in Chapter 4 are used as the main design criteria for this architecture. The first is the usability. This criterion advises that robotic system be designed so that implementation of new features and the extension of existing modules is relatively

7. SYSTEM DESCRIPTION

easy for developers, and facilitates maintainability. The second criterion describes the exchangeability of various components of the system. Often it is beneficial to have the option to exchange different parts of the system. For example, to use a part for different robots or to integrate another algorithm for a specific task. This exchangeability requires a reasonable deviation of functions between modules and flexible interfaces. This exchangeability and the usage of the SLA as a wrapper makes it easier to exchange a robot into another system because the social features do not have to be implemented again. The functional system can focus on the functional behavior. In this thesis, this system functions as a test bed for the DBCs, and was also used during the user studies.

7.1 Overview

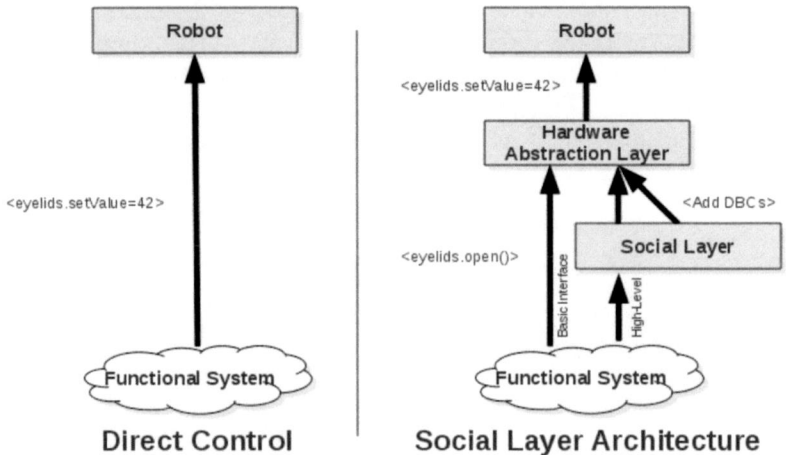

Figure 7.2: Comparing Social Layer Architecture (SLA) and direct communication. SLA enriches and interprets the high-level commands sent by the functional system and adds additional social behavior by using DBCs.

The main idea behind the Social Layer Architecture was to create a wrapper that encapsulates the robot and provides an interface that makes it possible to call higher level and basic functions instead of robot dependent low-level functions. In addition this wrapper contains a social layer that generates a basic social behavior using the DBCs. These DBCs only make use of robot independent behavior patterns that do not contain any direct joint commands. To interpret this pattern or other commands sent to the robot, the SLA contains a Hardware Abstraction Layer (HAL) that translates these high-level commands into specific robot motion. The HAL is the only module in the SLA that contains robot dependent code and functions as a driver. This makes it possible to use different robots with the same wrapper and interface. In Figure 7.2 the different strategies are presented. On the left side, the direct connection is shown. The functional layer in this case can be each robotic system, no matter how complex it is, that produces only functional behavior for a robot. This might be a system that works in a kitchen and stacks dishes or pours juice. At some point, this functional layer communicates with the robot by using robot dependent commands. For example `eyelids.setValue=42` is used to open the eyes. Of course many robotic architectures already have its own HAL that encapsulates the robot-specific commands from the system, but these drivers do not provide an additional social layer. The SLA in this case, functions as an alternative HAL that can be used instead of a functional driver without social features. In this thesis it is assumed that the functional system has no driver of its own and uses the SLA. The SLA is shown on th right side. By using this system, it is possible to use basic commands instead of robot-dependent commands. For example, commands like `eyelid.open()` could be used without having to know what command exactly opens the eyes. The high-level command is translated by the HAL into robot commands. In addition the social layer produces DBCs on its own that animate the robot in parallel to help it appear more lively by adding, for example breathing motion or blinking. Another function added are the high-level commands with which it is possible to enrich commands that are sent by the functional system to make them more social. For example, if the high-level command for speech is used, the social layer automatically enriches the speech synthesis by speech synchronous lip-movements. A detailed description of the general structure, and the containing modules can be found in section 7.2

Other features that are part of this system are some perception modules that provide high-level data generated using the robot sensors. For example, face detection and a saliency module are integrated. The gathered data are stored in a database so that it can be accessed by the whole system. The perception modules are described in section 7.7. For communication between modules and to sent commands to the SLA, a new markup lan-

7. SYSTEM DESCRIPTION

guage was created called PiRA-XML. This data structure is capable of defining complete animations, that include many different joints, in only one XML-File. This animation is interpreted in the HAL and translated into robot commands to generate animation in the robot. Using the PiRA-XML makes it possible to exchange different modules as long as they use PiRA-XML as their interface. How this markup language is defined and what commands can be generated is shown in section 7.3

The above mentioned features mainly focus on improving the exchangeability as mentioned in the fourth criterion for social robots. In terms of usability (third criterion for social robots) some tools are integrated to make it easier to maintain and use the SLA. First, the PiRA-XML is defined in a humanly understandable way. In addition, an animation editor is designed that makes it easy to generate animations by using a Graphical User Interface (GUI). This program, called PiRA-Editor, provides the option of defining whole body movements without directly programming the animations in plain text. The joint angles and the timing can be set for each joint by moving an anchor (little draggable boxes) on a time scale. Further program details and a usability study are presented in section 7.10. In terms of maintainability the strict division of different functions to discrete modules facilitates ease of use with this architecture from a developers perspective.

The modules that were developed for this robotic system were implemented in Java SE 7. For communication between modules the XCF [164] middleware was used, and for communication to the robot the middleware YARP [105] was used. The architecture itself was inspired by other robotic systems but contains several new approaches that come up by analyzing these. Chapter 4 describes the motivation for the SLA design. What follows is a detailed look at all the components, its functions and especially how they contribute to the two developer-centered criteria for social robots.

7.2 Systemstructure

In this section the architecture of the SLA system will be presented. At first, the interfaces that can be used by external functional systems will be described, followed by a detailed description of the implemented system. In the end, some sample test scenarios will be discussed to illustrate the underlying mechanisms. To understand the communication between the components, the middleware used, in this case XCF [164], needs to be described. XCF allows the exchange of information between different programs within one system. Thereby, it is irrelevant if the sender and receiver run on different computers within a network, or different operating systems, or are implemented with different programming languages. To establish communication, a server tool has to be running on which each XCF component has to register so that communication can be established. The information that are exchanged via XCF can be XML files or binary data. It provides three different types of communication. In `Publisher/Subscriber` mode, the publisher provides date such as the position of faces in a video stream every time it has computed new data. The subscriber can subscribe to this publisher and receive new data as soon as it is published. In the `Server/RemoteServer` mode, the server provides a function that can be executed by the remote server as needed. For example, the server provides the function to receive a new robot command and the remote server can call this method whenever it is needed. The third type of communication is the `ActiveMemory`. This is a type of database that can store data and provides access to all XCF components that register on the database.

Assuming the SLA as a black box, three different types of interfaces are visible for an external system. These are basic interface, high-level interface and the high-level data.

The basic interface provides functionality to send basic commands to the robot. These basic commands can be simple settings of joints into a certain state such as the angle ,triggering predefined animations such as shaking the head, or performing more complex whole body animations. Applying new states like joint angles or new facial expressions to a set of joints is called a posture that is executed on the robot. This postures have no execution time and only set one value to each joint which is applied as quickly as possible. The second possibility is to send an animation which defines a change over time for several joints in parallel. These basic commands can be sent to the SLA by using XCF. To send a command the sender first has to instantiate a `XCF::RemoteServert` to connect to the provided `XCF::Server` that is part of the basic interface. After establishing the connection the provided `setCommand` method can be used to send the PiRA-XML, containing the command, to the server for execution. The basic interface is capable of handling all commands and animations that can be defined in PiRA-XML which will be explained in section 7.3.

The high-level command is a set of functions that are provided beside the

7. SYSTEM DESCRIPTION

basic interface. These additional functions have their own interfaces, and can therefore be more complex and flexible. The only high-level command that is implemented at this point is the speech server that generates speech synchronous lip movement (see section 7.8).

The third interface type are the high-level data. These are data gathered by the robot ranging from simple records of executed commands to complex information such as the internal state of the robot or the position of faces in the field of view. All these data can be accessed externally using the high-level data interface.

Using these interfaces external programs have access to all necessary information concerning the robot, and the option to steer the robot without considering the social behavior that is automatically generated in parallel.

In the following the internal structure of the SLA will be explained. It consists of three layers. Combined with the functional layer, the system has four layer in total. Each layer is designed such that it has its dedicated tasks that create a semantic distinction between the layers. This design fulfills the first criterion of social robots because it facilitates maintainability and usability for developers. If there have to be changes in the social behavior only the social layer is effected, and nothing else. The four layers are the Functional Layer, Memory Layer, Hardware Layer and the Social Layer. Figure 7.3 illustrates all layers, modules and their connections. The light gray boxes represent semantic groups that do not necessarily have to be implemented in one module. The color of each arrow represents the connection type, and refers to the used middleware or other means of communication.

The *Functional Layer* has already been explained above, and represents the external system that uses the social robot. For this layer the three presented interfaces are visible, and can be used to control the robot.

The *Memory Layer* contains three databases that can store different internal data produced by the robot or define its internal states. These three databases are ActiveMemorys from the XCF middleware, and can be fully accessed by all components within the SLA as well as external components via the high-level data interface. The first database is the `PerceptionMemory` which contains all information collected by the sensors of the robot. This could be simple raw data or already interpreted high-level data such as the position of faces in the field of view and so forth. The `LongTermMemory` contains internal states of the robot like emotional state, interaction state (idle, interactive, ...), and long term memory. The

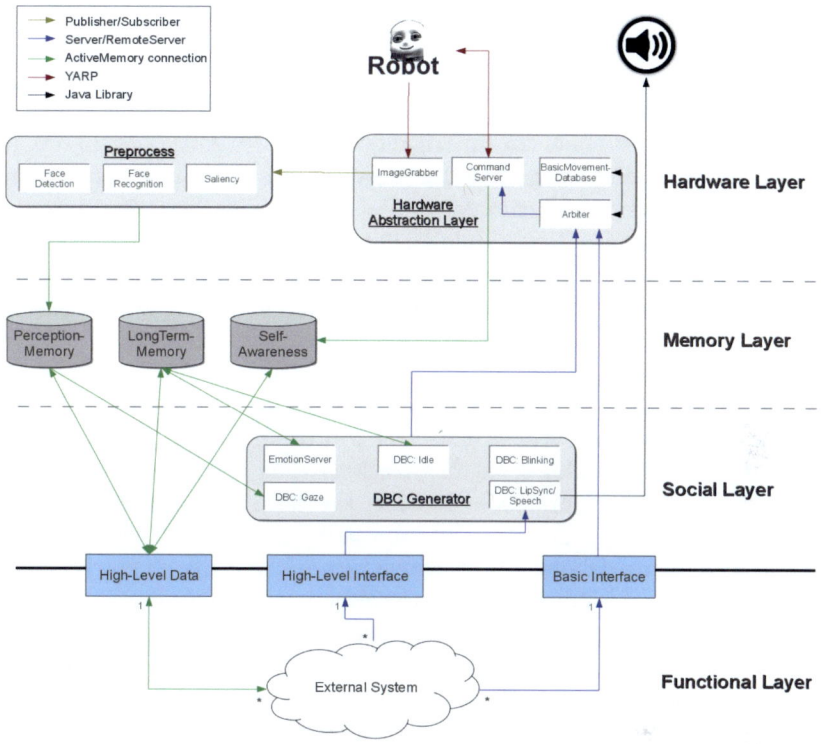

Figure 7.3: Block diagram illustrating the Social Layer Architecture

last type of memory is the `SelfAwareness` memory which contains a log of all movements carried out by the robot. The Memory Layer is explained in more detail in section 7.4

The *Hardware Layer* is composed by three different semantic groups. First the hardware itself is one part of this layer. All hardware components that are used and controlled by the SLA are in the hardware group. In this case, the robot iCub and a speaker are used since the robot does not have one. Additional sensors such as laser sensors or Vicon cameras would also be part of this group. The second group is the `HardwareAbstractionLayer`. These are the only components that have direct access to the hardware and provide interfaces for the rest of the system to use. When changing hardware components, this should be the only group that has to be

7. SYSTEM DESCRIPTION

adapted to integrate new hardware. This layer has four modules at the moment. The `Arbiter` checks parallel incoming commands on conflicts (see section 7.5), the `CommandServer` translates the PiRA-XML commands into robot commands (see section 7.6), the `BasicMovementDatabase` stores predefined animations that can be triggered by PiRA-XML commands (see section 7.5) and the `ImageGrabber` provides a stream of camera images that are captured by the cameras of the robot. The `ImageGrabber` is realized as an iceWing [91] module that grabs the image of the camera in the right eye of the iCub using YARP. Furthermore it provides this stream as `XCF::Publisher` so that it can be used by other modules all over the system. Within the SLA only the `Preprocess` uses this stream for further image processing. The task of `Preprocess` is to generate high-level data out of the provided sensor data. In the actual implementation, this is the robot's video stream. Within this group three modules are used. First, `FaceDetection` provides a set of positions that correspond to detected faces within the field of view. Second, the `Saliency` module computes the most interesting regions of the camera image using a salience module. Lastly, `FaceRecognition` can recognize persons that interacted previously with the robot. The `Preprocess` is described in detail in section 7.7

The fourth layer is the *Social Layer* that contains the social behavior of the robot and is the most important within this system. Here, most of the DBCs are generated that will be evaluated within this thesis. To generate these DBCs, four modules are implemented that realize gaze, idle movements, blinking and lip movements. All of these have access to all necessary data such as face positions and saliency and communicate with the robot by using PiRA-XML commands that are sent to the `Arbiter`. This behavior is produced continuously unless the `Arbiter` decides to suppress it when commands with a higher priority, for example by the `Functional Layer`, have to be executed. A detailed description of the `Social Layer` is provided in section 7.8.

As stated above, using the layer structure has its advantages especially in terms of maintainability and exchangeability. Nevertheless there are some disadvantages that will be discussed here. Often it is challenging to divide the functional and the background behavior onto two different modules. In most cases it would be more efficient, on an implementation level, to generate both behaviors in the same module where each behavior is aware of the other. When dividing behaviors such that they do not know what the other behavior is doing additional algorithms have to be implemented to address with conflicts and interferences. For example, if gaze were to be realized within one module, the functional behavior could actively turn off the gaze DBC or wait until it was possible to perform a gaze shift. By dividing the behaviors, an additional conflict detection has to be implemented that solves these problems. Often systems that fully integrate the background functions into the functional system are more robust and have advantages in speed. But they also lack flexibility. By using the proposed layer structure, it is possible to create a fully functional social robot with an attachable functional system. The creation of such an architecture aims to be a final product deliverable to other researchers who want to use a self-contained social robot that can be attached to their functional system.

7.2.1 Test Cases

To better understand the internal information flow, two test cases will be described. These two cases were chosen because they represent the most common communications that occur within the system, and cover about 90% of the communications that are carried out. For each test case, a sequence diagram of one exemplary run that illustrates it best is given along with its function test. In addition, for one test case, the execution time was evaluated to see how long a command needed to be executed on the robot. *Test Case 1* describes an external request to execute an animation on the robot while DBCs are produced in parallel. The problem in this case is that the DBCs and the animation use the same joints which would typically create conflicts. In case of a conflict, the animation has higher priority, and the DBC is to be suppressed. To keep things simple, the test case is explained based on the gaze DBC and a nodding behavior that is triggered externally. The complete sequence is shown in Figure D.1. So as to illustrate the execution of an animation the gaze cue is left out for a moment. The animation is triggered by an external source (:Functional Layer). This source sends the PiRA-XML file containing the animation to the arbiter by calling the `setCommand` method of the `XCF::Server`. This method is part of the basic interface and is therefore accessible to external sources. After receiving the animation, the `Arbiter` checks to see is there are any conflicts with already active commands on the robot, and none are found in this example. After this test, the `Arbiter` forwards the command to the `CommandServer` via `XCF::Server` that executes the animation on the robot. The `CommandServer` and the robot are presented

7. SYSTEM DESCRIPTION

as one lifeline in the sequence diagram. The algorithm that is used by the `CommandServer` to execute a command on the robot is described in section 7.6 and is therefore excluded here. Depending on what command is sent, the method called on the `CommandServer` returns immediately, for a posture, or waits if there is an animation until it is finished. In this example, an animation is sent and the execution lasts a certain period of time before the resources are freed. After executing the animation, the `Arbiter` returns and informs the sender that the command was accepted and that the execution was successful. If there is an error while executing the command, different return values can be used to inform the sender about it. This is the case when another process sends a command to the robot while an animation is still running. To illustrate this, it is assumed that the animation that has been sent by the functional system is still running. During this animation the gaze DBS sends the command to perform a gaze shift. After sending it to the `Arbiter`, this command is checked for any conflicts. If a conflict occurs, the whole command is rejected. In this example, the nodding animation and the gaze shift use the same joints in the neck which causes a conflict. Because of this, the method terminates immediately and informs the sender that the command caused a conflict and was rejected. The sender now has the option to decide what to do next.

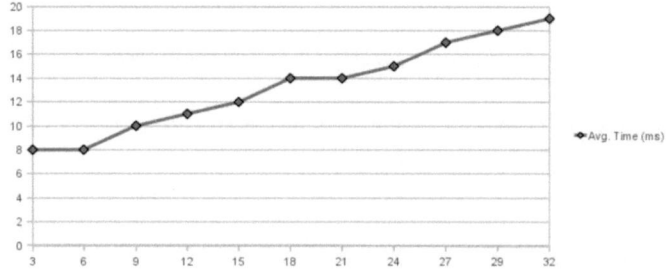

Figure 7.4: Execution time needed to execute a command in the SLA. Duration is shown on the Y-Axis. The X-Axis represents the complexity of the command

To test the function of this test case, a set of animations was created that lasted at least 5 Seconds. During this timespan, a set of postures and other animations were sent to the robot. The two used sets were chosen such that some caused conflicts and others could exist in parallel, and cover all possible combinations. For example, nodding and a leg movement can be executed in parallel. During the test, the arbiter always decided as predicted and passes the function test. Another test was carried out to see how long a single command needed to be executed. Therefore a small program was written to send 100 posture commands to the robot one after the other. The mean execution time for all 100 trials was computed and taken as a measure. The commands sent were basic posture commands with different complexity. The first 100 commands set new positions for one joint, the second 100, two joints and so on. Figure 7.4 shows the mean durations plotted that were needed to execute one command. The X-Axis represents the complexity, and the values stand for the number of joints that were used. The execution time ranged from 8 ms for one joint up to 19 ms for 32 joints with a linear increase in execution time. This duration is short enough to execute social behavior on the robot without noticeable delays for the human user, even with many joints. This value only measures the time that is needed to send the command to the robot, not the time the robot needs to set the joint angles. That depends on the velocity that is set for each joint, and is not relevant for this test.

The *Test Case 2* describes the realization of the gaze DBC described in section 6.4 starting from the face detection up to the final generation of the gaze shift in the robot. The main details of this communication are shown in a sequence diagram in the Appendix (see Figure D.2). First the necessary data had to be produced. Therefore the `FaceDetection` and the `Saliency` module continuously analyzed the video stream and exported these data to `PerceptionMemory`. This process is represented by the continuous `Gather Face Data` or `Gather Saliency Data` in the diagram. These data are relevant until new data are computed and added to `ActiveMemory`. Whenever the gaze DBC needs the position of a face or a salient region these data can be requested from the database (represented by the `Request Face Position()` or `Request Saliency Data()`). In this example, the DBC first requests a face position and computes the gazing direction based on this data. The command to shift the gaze is translated into a PiRA-XML file and sent to the `Arbiter`. This component either executes the command or rejects it because of conflicts. After finishing the first command, the gaze DBC generates the next gaze shift to a salient region. The procedure is nearly the same as for faces. The only difference is that the necessary date are gathered from the `Saliency` module.

To test the functionality of the test case, some long term tests have been

7. SYSTEM DESCRIPTION

carried out to check if the system crashes after a while, or produces errors during execution. The system ran in background for several hours without producing any problems, errors or crashes. During these tests, all DBCs were running, including image preprocessing and command executions on the simulated iCub.

7.3 PiRA-XML

Communication within the SLA and its external components is mainly realized with the PiRA-XML format. In this section, this data format is described and its advantages and disadvantages are discussed. PiRA stands for Platform independent Robot Animation, and is a markup language to represent whole body animations usable across different types of robot. It is also possible to have unrestricted movements timings for each single joint as when using key frames for the animation. This data format fulfills the developer-oriented criterion by being exchangeable and easy to use. First, the XML structure is readable on many operating system, and is robot-independent. Second the structure is human-readable and can be edited with a plain text editor, or with the PiRA-Editor. In first place, this structure is a container for the defined command. Its header stores various types of information about the command such as the type of robot, whether the command is an animation or a posture, and the duration of the animation, if needed. The embedded command can define different types of commands. First, a posture is a command that sets a single joint angle for several joints so that the robot can move to the defined posture. This command has no duration, the robot simply moves its joints as quickly as possible. The animations that can be created, animate the movements of the robot over a period of time. Here it can be defined that a certain joint should reach a position in a specific point in time, and then move to another position while the other joints execute similar commands by using other timings for the goal positions. This makes it possible to animate a robot more flexibly instead of just applying one posture after the other as with key frame animations. Furthermore, it is possible to trigger predefined animations that are stored in a robot dependent library. These could be a nodding, wave or idle animation.

Other similar formats have been developed over the past few years. The most prominent animation structure is the Behavior Markup Language (BML) [149]. The BML is an extensive structure that allows the user to define robot animations containing speech and synchronized movements. The BML does not define the exact animations, but instead uses external formats to replay an animation to an exact point in time and/or duration. One example is MURML [86] which can be used as another XML based animation markup language. The language focuses mainly on defining gestures that are presented with the right timing synchronized to speech. The advantage of using PiRA is that this language can define any animation by giving the user full control of each joint, represented as individual streams, so as to generate very natural movements. One disadvantage compared to MURML is that the PiRA animations cannot be parameterized and can only define a fixed movement after having been designed once. Adding speech is also possible with PiRA, as described in section 7.8, but the BML provides more flexible ways to integrate animations with speech. The focus of PiRA lies in defining natural animations and not in synchronizing speech and behavior or defining a whole sequence of animations. An optimal combination for future implementations would perhaps be to use BML as a container for communication especially for speech and integrate the animations into BML using PiRA-XML.

Often robot animations are generated in two different ways. First the motion is designed manually by defining key frames that define fixed postures and the robot moves from one posture to the next at the right time. Often these key frames lead to unnatural movements because it is obvious that all joints used change direction, or start or stop at the same time (referring to the fifth principle of animation). On the other hand, whole movements can be recorded by using motion capture systems, or moving the robot to save all joint angles during the complete animation. Although these animations look very natural, they are difficult to design manually without recording them. In addition saving every joint angle to every point in time needs a relatively large amount of storage. That said, the naturalness of the key frame animations can be improved by adding more key frames, which would more closely approximate the recorded animation. The disadvantage is that this increases the complexity and amount of data, lowering the advantages towards the recorded animation. This raises the question of building a structure that combined the natural movement of a recorded animation with the simplicity of key frame animations. One way to achieve this is to have an individual definition for each joint that contains a sequence of two parameters for simple joint control: time and angle. This tuple, which is called *anchor*, defines when the joint should

7. SYSTEM DESCRIPTION

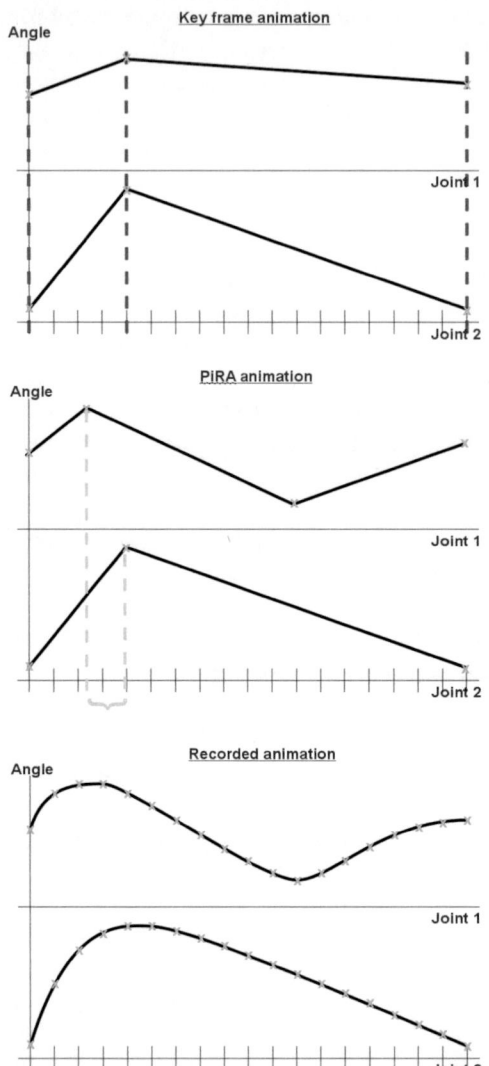

Figure 7.5: Three different animation strategies to encode the same movement. Each green "X" represents one data point that is needed to define the animation. Key frames are represented as red lines, which are the only point in time where data points can be placed for all joints. The two gray lines illustrate that there are no restraints in the timing for the PiRA-XML and that the anchors could be placed anywhere.

have a certain angle. A sequence of tuples defines the motion of a joint, which combined with all other sequences of tuples defines the whole body animation. This makes this approach very flexible compared to the other two because it can be changed very easily without rerecording the entire sequence, and while nevertheless having the freedom to define anchors at any point in time without being restricted to key frames. In addition, in most cases the code is shorter for the same animation compared to both other strategies as most information do not need to be repeated. In addition, the PiRA format uses interpolations to generate the values between each anchor, which further decreases the amount of code that has to be stored. In Figure 7.5 all three strategies are presented. In this Figure, in all cases the same movement is encoded, whereas the recorded animation represents the best version of the movement. For the key frame animation, it is noticeable that for the first joint the second extrema cannot be represented because there is no key frame. In addition, the first extrema for both joints have the same timing, which is a deviation from the original motion and might be noticeable by a human user. The PiRA animation provides the flexibility to represent all important aspects of the animation by using only seven data tuples.

7.3.1 The structure of PiRA commands

In the following section the structure of a PiRA-XML command is presented. This structure is very flexible and allows the combination of different types of low-level and high-level joints in only one command. Figure 7.6 shows the general structure of an XML file. It can a posture command that has no length and just applies the given set of joint positions to the robot. The speed with which the movements are carried out depends on the previously defined acceleration of the joint. Further, it can also be an animation with a given duration that defines the movements of the robot over a certain time. In both cases, use of different joint types can be made where several can be used in parallel. Postures as well as animations can use `PostureJoints` that define a whole group of predefined joint positions to set common postures with only one command, or `SingleJoints` like `SimpleJoints` or `CustomJoints` that both define the configuration of a single joint. In addition, postures can use predefined animations from the `BasicMovementDatabase` (BMD) which contains animations such as nodding, waving or idle motions. BMD commands cannot be used within animations. After sending the command, all subcommands (including

7. SYSTEM DESCRIPTION

PostureJoints and BMD) are translated into a set of SingleJoints that can be executed by the robot.
A typical PiRA-XML command may look like the following:

```
<PiRAxml robot="iCub" robotType="humanoid" type="posture">
      <Joint jointType="SimpleJoint" name="leftarm_shoulder_pitch">
            <Anchor value="-47" />
      </Joint>
      <Joint jointType="PostureJoint" name="FaceExpression">
            <Anchor value="Angry" />
      </Joint>
</PiRAxml>
```

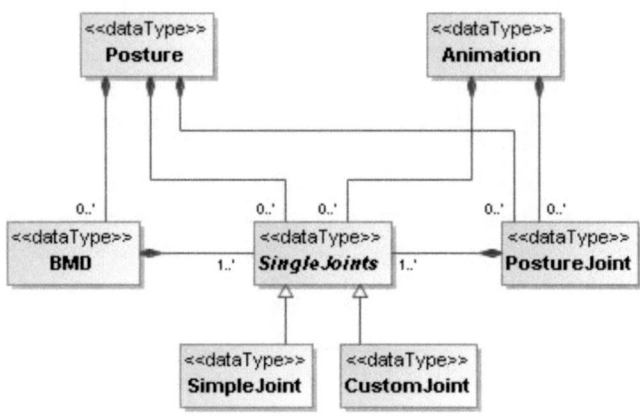

Figure 7.6: PiRA data structure as UML Class Diagram

This command is a posture as can be seen in the `type` attribute of the PiRA-XML tag. The attribute `robot` defines the robot for which this command was originally designed. The attribute `robotType` defines for which type of robot this command is usable. If other types of robots are used, such as a four legged pet robot, they could require other joint configurations in the event they have different DOF. This command sets the pitch of the left shoulder to a value of -47 degrees and displays a facial expression of anger. Therefore the first is a SimpleJoint that only controls one joint and the second a PostureJoint that controls all joints used to control facial expressions such as lips, eyebrows and eyes.

The header of this XML file has the following structure as a Relax NG [35] schema:

```
<element name="PiRAxml" xmlns="http://relaxng.org/ns/structure/1.0">
        <attribute name="robot">
                <text/>
        </attribute>
        <attribute name="robotType">
                <text/>
        </attribute>
/* -- Define command type ----------------------*/
        <choice>
                <attribute name="type">
                        posture                 /* first command type */
                </attribute>
                <group>
                        <attribute name="type">
                                animation       /* second command type */
                        </attribute>
                        <attribute name="duration">
                                <text/>         /* length of the animation */
                        </attribute>
                        <optional>
                                <attribute name="StepWidth">
                                        <text/> /* time between two animation steps */
                                </attribute>
                        </optional>
                </group>
        </choice>
/* -- Define command--------------------------*/
        <oneOrMore>
                <element name="Joint">
                        <choice>
                                <attribute name="jointType">
                                        SimpleJoint
                                </attribute>
                                <attribute name="jointType">
                                        CustomJoint
                                </attribute>
                                <attribute name="jointType">
                                        PostureJoint
                                </attribute>
                                <attribute name="jointType">
                                        BMD
                                </attribute>
                        </choice>
                        ...                     /*animation/posture has to be defined here*/
                </element>
        </oneOrMore>
/* ---------------------------------------------*/
</element>
```

This shows that these commands can be either postures or animations. If they are animations, a duration needs to to be added. Depending on the type of command, the structure of the `Anchors` differs. As elements, any amount of joints of a chosen type can be added. First the description confines on posture commands. Animations will be considered later.
The `SimpleJoint` and the `CustomJoint` belong to a group called `SingleJoints`. These joints have in common that they only control one single joint. The `SimpleJoints` can control classical joints, like an elbow, that have one

7. SYSTEM DESCRIPTION

parameter i.e. the angle. The `CustomJoint` can be used to control joints that use more than one parameter. For these joints, in most cases an individual `CustomJoint` has to be defined. In the actual system, gaze uses such a joint. Two parameters, X and Y, are used to set the coordinates the robot should look at. For both joint types, the only information needed is the name of the joint that has to be controlled. An additional tag called `Anchor` contains the parameter that should be set. The Relax NG schema for a general `SimpleJoint` and the `CustomJoint` for gazing is the following:

```
...
<choice>
/*----------SimpleJoint------------*/
        <element name="Joint">
            <attribute name="jointType">
                SimpleJoint
            </attribute>
            <attribute name="name">
                    <text\>              /* Name of joint */
            </attribute>
            <element name="Anchor">
                <attribute name="value">
                    <text\>              /* Value for the chosen joint */
                </attribute>
            </element>
        </element>
/*----------CustomJoint------------*/
        <element name="Joint">
            <attribute name="jointType">
                CustomJoint
            </attribute>
            <attribute name="name">
                gaze
            </attribute>
            <element name="Anchor">
                <attribute name="x">
                    <text\>              /* x coordinate for gazing */
                </attribute>
                <attribute name="y">
                    <text\>              /* y coordinate for gazing */
                </attribute>
            </element>
        </element>
/*--------------------------------*/
</choice>
...
```

The remaining joint types `PostureJoint` and `BMD` can control multiple joints at the same time. The general strategy behind these joints is to provide the option to trigger postures and animations on the robot by using high-level commands without having to take into account the exact control of the robot. All robot dependent joint configurations and animation files are stored in the `BasicMovementDatabase`. The PiRA-XML only needs to trigger a "nod" and the BMD translates this into the necessary commands, which can differ for different robots. This BMD can store commonly used animations and postured in one place. The animations can be optimized for each robot, although for the `SocialLayer` or `FunctionalLayer` the same commands are always available to wrap the underlying robot control. The structure of both commands is very similar. The main difference is that the BMD commands are not allowed in an animation because these commands are an animation in itself, and therefore cannot be combined with other animations. Analogous to the `SingleJoints`, the name and the corresponding `jointType` have to be set for this command. The BMD command does not need any additional parameter, in contrast to the `PostureJoints` where an `Anchor` tag has to be defined for posture parameters, for example facial expression when using the `FaceExpression` joint. In the following the Relax NG schema for these two joint types is shown:

```
...
<choice>
/*----------BMD------------*/
        <element name="Joint">
                <attribute name="jointType">
                        BMD
                </attribute>
                <attribute name="name">
                        <text\>                  /* Name of triggered animation */
                </attribute>
        </element>
/*----------PostureJoint------------*/
        <element name="Joint">
                <attribute name="jointType">
                        PostureJoint
                </attribute>
                <attribute name="name">
                        <text\>                  /* Name of PostureJoint */
                </attribute>
                <element name="Anchor">
                        <attribute name="value">
                                <text\>          /* Parameter for this joint */
                        </attribute>
                </element>
        </element>
/*--------------------------------*/
</choice>
...
```

To use animations, a few simple modifications need to be made in the structure of the joints. For animations, all joint types are allowed except BMD joints. When defining a posture, all used joints need to provide a single `Anchor` which contains the parameter that is set when the command is applied. For animations, a list of `Anchors` has to be defined that represent the change of this joint over time. Each `Anchor`, therefore gets an additional attribute that defines the time in milliseconds when the value of this `Anchor` should to be reached. At least two `Anchors` are needed. One to define the start value and one for the value at the end of the animation. In between, any amount of additional anchors can be defined. When the animation is executed, the `CommandServer` streams the animation to

7. SYSTEM DESCRIPTION

the robot by generating `PostureJoints` that contain the posture of one specific time in the animation (e.g.: 200 ms). After a short time, which can be configured by setting the `StepWidth` parameter, a new posture is generated and sent to the robot (e.g: 240 ms). If the `StepWidth` is not set, the default is 40 ms. To generate the values between two `Anchors`, these are interpolated for `SimpleJoints` which generates a smooth animation. `PostureJoints` use mostly discrete values for their parameters, as for facial expressions, which can not be interpolated. In this case, the new value is set at the timing of the `Anchor`. A sample animation for one joint could look like the following:

```
<PiRAxml robot="iCub" robotType="humanoid" type="animation" duration="2000" StepWidth=\"30\">
    <Joint jointType="SimpleJoint" name="leftarm_shoulder_pitch">
        <Anchor time=0 value="-47" />       /* First anchor. Time must be set to 0 */
        <Anchor time=1200 value="30" />     /* All anchors have to be sorted by time */
        <Anchor time=2000 value="10" />     /* Last anchor. Defines the end of the
                                               animation */
    </Joint>
</PiRAxml>
```

Compared to key frames and recorded animations, the PiRA-XML has some advantages that allow it to generate animations more ease, flexible and with less lines of code. Given its actual functionality, the PiRA-XML can only be used with humanoid robots, although extending it to other types of robots is possible. For example a `CommandServer` for the Sony AIBO exists that can be used to control this pet-like robot using PiRA-XML. However this server does not yet support `BMD`. Another planned extension is the creation of another `CustomJoint` to control the end-effector position of the hand with 3D coordinates. For this thesis this functionality was not necessary because manually designed arm movements are often more natural than those generated by a kinematic algorithm.

7.4 Memory Layer

The function of the `MemoryLayer` is to store and provide data necessary to the system. This layer can be thought of as the social robot's memory center. Three databases are included in the `MemoryLayer`. These are `PerceptionMemory`, `SelfAwareness` and `LongTermMemory`. Each of these databases has its own functions and represents different parts of memory as will be explained below.

All databases in this layer are realized using `ActiveMemorys`. This is a database included in the XCF middleware framework, making it is possible to communicate with the database using XCF. The data stored in the database are XML files that store, for example, PiRA-XML files. XCF clients send requests to use the stored data. It is possible to specify the XML data to be returned by using criteria match the desired XML file so that only the needed data are provided. These criteria might be the name of an XML data tag, such as "PiRAXML", or the value of an attribute or a certain XML structure. They are used as a filter for all stored data. Furthermore it is possible to register an XCF client on events as ad, delete or modify of an entry allowing it to inform components when new data are available. For example, the social layer is informed when there are new face positions for the interaction partner computed, to adjust the gaze of the robot. Lastly, it is possible to insert data into the database. This could be either a new entry or the modification of an existing one. All three databases are visible to all components of the system via the high-level data interface.

`PerceptionMemory` contains all data that were gathered and computed by the sensors. Its main function is to realize storage for the robot's perceptual process. Data that are computed during the `Preprocess` are stored within this database by adding XML data containing this information. For new information, a new dataset is created so that past data remains available as needed. In the actual implementation the position of detected faces and the results provided by the salience module are stored.

`SelfAwareness` memory contains information about the actions of the robot itself. All commands sent to the `CommandServer` are stored in this database so that the robot's past behavior can be restored.

`LongTermMemory` consists of all data relevant in the long term, or data structures that define the internal state of the robot. Long term data could be data about a certain situation or interaction, including human participants, time, location, etc. or information gathered from the environment such as labels for recognized objects or persons. No long term data are used for the SLA layer itself at the moment although this functionality is provided mainly for the `FunctionalLayer`. The second function of this memory is to store the internal state of the robot. It is stored within one XML structure that is updated when the internal state changes, meaning that only one structure exists at a time. This XML data looks like the following:

```
<InternalState>
    <EmotionState value="happy"/>
    <FeedbackMode value="idle"/>
    <DBCMode value="fullDBC"/>
    <ModuleActivation>
        <EmotionServer value="active"/>
```

7. SYSTEM DESCRIPTION

```
            <SpeechServer value="active"/>
        </ModuleActivation>
</InternalState>
```

This XML defines several states of the robot including emotional state, the feedback mode the robot is in and which DBCs are active. Within the tag `ModuleActivation` different modules can be made active or inactive. Using this, it is possible to activate or deactivate certain functionality or modules from anywhere in the system. By using the tag `DBCMode`, different DBC settings can be set. For example "fullDBC" is used to turn on all DBCs or "noDBC" to turn them all off. In some cases, this might be useful, for example, if no idle movements are to be carried out during a grasping task. Feedback mode and Emotional state can also be set here. The feedback mode defines the general mode the robot is in, i.e. "idle" when no interaction takes place or "interaction" if the robot interacts with another person. The feedback mode can be used to trigger different behaviors during the different states. For example, the robot does not have to search for faces if there is no face-to-face interaction. Emotional state represents the emotion the robot should display at any given moment. It can be used in a similar way. At the moment, the facial expressions of the robot are displayed depending on its emotional state. During an interaction, a module in the `FunctionalLayer` might influence the emotional state of the robot. I.e. if a task did not go well the robot will become "sad". This module simply has to change the emotional state in `LongTermMemory` and the DBCs can react automatically to this change by varying their behavior. The internal state is a centralized robot configuration that can be influenced by the whole system and thereby influences other parts of the system.

7.5 Hardware Abstraction Layer: Arbiter

The `Arbiter` is a component that is necessary to prevent conflicting commands. When designing a system with an interface for a `FunctionalLayer`, a component is needed to judge which commands can be carried out and which cannot because the SLA has no control over which commands are sent to the robot. For example, two commands can be sent by the functional system to turn the head in different directions while the gaze DBC tries to do the same. A decision must be made as to which command has the highest priority, and which need to be postponed. In addition, the sender of rejected commands needs to be kept apprised. The arbitration itself is a very complex task and does not lie in the scope of this thesis. Consequently, a simplified arbitration is implemented that follows a "first come, first serve" strategy. The implemented `Arbiter` consists of two components. The first is the `Arbiter` itself. It can receive commands by `XCF::Server` which are checked for conflicts and forwarded to the `CommandServer` for execution if they are accepted. In both cases the Arbiter informs the sender of the command about its status. The second is the `BasicMovementsDatabase` (BMD). This module is part of the `Arbiter` and checks if a received command contains any `PostureJoint` or BMD command. If this is true, it converts these commands into `SingleJoints` that replace the original to be executable by the `CommandServer`.

First, after receiving a command it is forwarded to the BMD module. PiRA-XML files can contain high-level commands such as postures and animations that are triggered by using keywords. These commands are not interpretable by the arbiter or the command server because these two components can only handle `SingleJoints`. Therefore, the first step after receiving a command is to translate all commands into `SingleJoints` as necessary.
Thereby the already existing `SimpleJoints` may need to be adjusted. This is the case if a BMD command is translated into an animation. Parallel `SingleJoints` then have to be translated into animations, as well. These animations have to set the same value for the joint during the whole animation to display the same movement as setting the value once.
In the case of a `PostureJoint`, this joint is replaced by the `SingleJoints` that define the triggered posture. These define the same posture as the original PiRA-XML file and can be forwarded to the arbiter. The described conversion of the command is the task of the BMD module. This mod-

7. SYSTEM DESCRIPTION

ule makes it possible to define frequently used postures (`PostureJoint`) and animations (BMD command) individually for each robot. The following examples show how the conversion of the commands is done. This command defines a posture that controls the facial expressions:

```
<PiRAxml robot="iCub" robotType="humanoid" type="posture">
    <Joint jointType="PostureJoint" name="FaceExpression">
        <Anchor value="Sad" />
    </Joint>
</PiRAxml>
```

Here the keyword "FaceExpression" is used in combination with a parameter in the `Anchor` tag that defines the facial expression. The command above is converted into the following command:

```
<PiRAxml robot="iCub" robotType="humanoid" type="posture">
    <Joint jointType="SimpleJoint" name="Mouth">
        <Anchor value="0B" />
    </Joint>
    <Joint jointType="SimpleJoint" name="RightEyebrow">
        <Anchor value="02" />
    </Joint>
    <Joint jointType="SimpleJoint" name="LeftEyebrow">
        <Anchor value="02" />
    </Joint>
</PiRAxml>
```

After converting this command, the `PostureJoint` is replaced by a set of `SimpleJoints` that can be interpreted by the `Arbiter`. In the second example a posture joint is defined such that it contains a BMD command that is set parallel to a `SimpleJoint`:

```
<PiRAxml robot="iCub" robotType="humanoid" type="posture">
    <Joint jointType="SimpleJoint" name="leftleg_knee">
        <Anchor value="-20" />
    </Joint>
    <Joint jointType="BMD" name="Stretch" />
</PiRAxml>
```

Here it can be seen that even though the BMD command is an animation, the command is defined as a posture. The name "Stretch" defines the animation that has to be carried out and needs no `Anchor` to define further parameters. The `SimpleJoint` sets the angle of the left knee to -20 degree. The command above is converted into the following command:

```
<PiRAxml robot="iCub" robotType="humanoid" type="animation" duration="4000">
    <Joint jointType="SimpleJoint" name="leftleg_knee">
        <Anchor value="-20" time="0" />
    </Joint>
    <Joint name="leftarm_shoulder_roll" jointType="Joint">
        <Anchor time="0" value="15"/>
        <Anchor time="655" value="20"/>
        <Anchor time="2838" value="20"/>
        <Anchor time="4000" value="15"/>
```

```xml
        </Joint>
            .
            .
            .
        <Joint name="torso_torso_roll" jointType="Joint">
            <Anchor time="0" value="0" />
            <Anchor time="760" value="2"/>
            <Anchor time="3209" value="-2"/>
            <Anchor time="4000" value="0"/>
        </Joint>
        <Joint jointType="SimpleJoint" name="Eye_Lids">
            <Anchor time="0" value="4A" />
            <Anchor time="1166" value="2F" />
            <Anchor time="3500" value="4A" />
            <Anchor time="4000" value="4A" />
        </Joint>
</PiRAxml>
```

The most important modification when converting a BMD is the change of the command type to an animation. The length of the animation is set to the length of the BMD animation. Furthermore, the BMD is replaced by a set of animations for each joint. The **SimpleJoint** for the knee has to be modified because it is not an animation in the original command. To achieve this, the **Anchor** is extended with a **time** attribute that is set to zero, which functions as a start point for the animation. An **Anchor** for the end of the animation is not needed because it is assumed that the **Anchor** with the highest **time** value defines the state that should be set until the end of the animation.

After converting the commands, the **Arbiter** distinguishes between posture commands and animations. If an animation is sent to the robot, all joints are blocked that are involved in the animation. To achieve this, the arbiter holds a list that contains the availability of all joints. For each command all used joints are extracted and compared with the list. If at least one joint is not available, the command is rejected. If all joints are available they are blocked and no other animation or posture can send commands containing the blocked joints until the first animation is finished. If both commands use different joints both are executed in parallel. In the Appendix Figure D.1 a sample execution of conflicting commands is shown as a sequence diagram. Posture commands do not block any joints because they do not have a timespan. If another command uses the same joints the new command overwrites the first one. In addition it is possible to block joints without using them through a special PiRA-XML command. This can be useful when it is necessary for the robot to keep certain joints still, as when a steady head is needed to gather camera data. In this case, the torso and head joints can be blocked to not allow other processes to control them.

7. SYSTEM DESCRIPTION

All DBC patterns are realized as posture commands that are sent to the `Arbiter`. The fact that posture joints do not block joints but are rejected when they use blocked joints, generates a kind of priority. They are only produced if the joints are not used, and they do not block any resource that hinder the `FunctionalLayer` when executing commands. This suppresses the execution of DBCs when it is not needed and displays it if no other command is executed.

7.6 Hardware Abstraction Layer: Command server

The `CommandServer` is the direct interface of the robot. This module consists mostly of robot dependent components, and translates the PiRA commands into robot commands. Its interface is an `XCF::Server` but can only interpret PiRA-XML files containing `SimpleJoints` and `CustomJoints`. This is not a problem because the other layers do not communicate directly with the `CommandServer`. The `Arbiter` does all the pre-processing to translate all `BMDs` and `PostureJoints` into `SingleJoints` and checks for conflicts. The task of the `CommandServer` is only to execute `SimpleJoints` and `CustomJoints` on the robot. The output of this module are YARP commands that apply the given values to the joints of the robot.

The internal structure of this module is the following. A dedicated component is responsible for receiving all commands that are sent to the robot. After receiving them, the commands are classified as a posture or an animation command, a key step as they need to be executed differently. If the command is a posture command, the `SetCommand` component divides the command into smaller ones that contain the instructions for only one `SimpleJoint` or `CustomJoint`. These commands are distributed to the corresponding `JointGroup` that is able to execute this specific command on the robot by using YARP. If the command is an animation the procedure is slightly different. The `StreamAnimation` realizes the streaming of animations to the robot. Therefore the animation is segmented into small sections which have by default a length of 40 ms. These segments are send to the `SetCommand` component as posture commands one after the other to generate a smooth animation. A class diagram of the structure is shown in the Appendix Figure D.4

Some translation needs to be done before sending a posture command to the robot. Since the commands are robot independent, a mapping to the robot command has to be carried out. For the iCub this is done using the `YARP_IDMap`. This is a HashMap that returns the necessary information needed for the translation. If, for example, the left elbow should be moved, the PiRA-XML uses the keyword "leftarm_elbow" as a key for the HashMap. The returned information contains the name of the responsible `JointGroup`, that then realizes the execution, the robot dependent name of this joint and the mapping of PiRA-XML value to robot value. Depending on the robot, the given values for the joints may differ. For example the position of the elbow is defined as 0 degree for the fully stretched elbow, and 90 degree when the elbow is at a right angle. If the robot values do not fit this definition, they need to be converted. This transformation is done in the `SetCommand` component. After converting the joint information, the command is divided into subcommands that only contain one joint. Depending on the corresponding `JointGroup`, the commands are forwarded here. For example, a gazing command is sent to the `JointGroup_Gaze` and the elbow command is sent to the `JointGroup_LeftArm`. At this stage it is irrelevant if this joint is a `SimpleJoint` or a `CustonJoint`. The interpretation of the anchors is done within the `JointGroups`

To execute an animation an additional step needs to be carried out before sending it to the robot. The `StreamAnimation` component generates a stream of posture commands out of the given animation that are applied automatically one after the other in the robot. To achieve this this module analyses the anchors of the animation and exports the robot posture, containing all joints, for the specific time of the animation, for example at 3200 ms, by interpolating the values between the two anchors. This posture command is sent to the `SetCommand` component that executes this posture on the robot. After a certain time, the `StreamAnimation` generates the next posture command for the time 3240 ms and sends this posture to the robot. Displaying all states of the animation one after the other in the robot generates an animation.

Simply applying the postures to the robot does not generate a smooth animation. Two problems can occur when just sending one posture after the other. The example on Figure 7.8 illustrates this issue for a sample joint. The reason for these problems lies in the predefined acceleration of the joints. If the joint needs to perform a movement to reach a 30 degree

7. SYSTEM DESCRIPTION

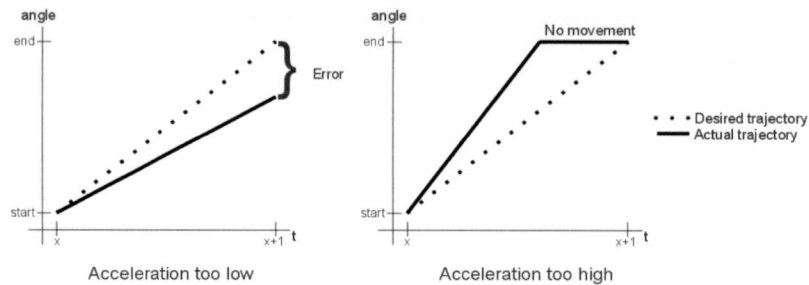

Figure 7.7: Joint movements depend on acceleration. If acceleration is too low, the desired angle may not be reached; if it's too high, the angle is reached too early.

angle after 200 ms, and this command is sent to the robot without any regulation in most cases the joint will reach the desired value too early or too late. Especially when using very short steps, this effect becomes more and more important. If the general acceleration is too high, the joint reaches the next posture too early and stops moving until the next posture has to be carried out a few milliseconds later. This causes shaky motion which should to be avoided. If the acceleration is too slow, the movement itself is smooth because the joint is still moving when receiving the new value and can continue. But the movement will never reach the desired value in time. Another factor to be considered is that the dynamics of the robot may change. The acceleration of the shoulder needs to be much higher when the elbow is stretched than when it is bend. In the worst case, the elbow joint moves during the shoulder movement, changing the dynamics while the arm is moving. This makes it necessary to regulate the acceleration of the joint dynamically during the movement. Before each animation step, the acceleration of each joint is adjusted depending on its actual angle. If the joint moves too fast the velocity is reduced. If it is too slow, it is increased. This guarantees that the whole movement is carried out without being shaky while also reaching the desired value in time. All the regulation is carried out within the `JointGroups`. The `StreamAnimation` only generates the next posture and provides the time when the next posture will follow, which will be needed to compute the correct accelerations for the joints.

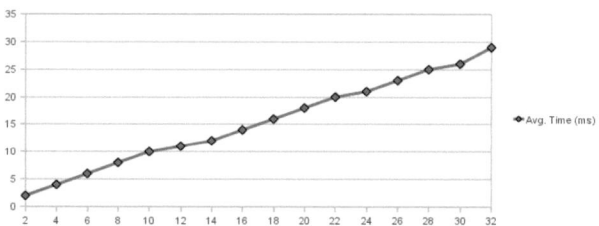

Figure 7.8: Execution time needed to generate a posture during an animation. Duration is shown on the Y-Axis. The X-Axis represents the complexity of the command

The animation streaming uses one parameter that can be set within the PiRA-XML command. This parameter is called StepWidth and defines the time between two postures while streaming the animation. This parameter can be set for the whole animation by using the StepWidth attribute in the PiRA-XML tag. Otherwise the default value is set to 40 ms. A higher value for this parameter results in less traffic to the robot but also makes the animation less fluent. Furthermore, some details of the animation may vanish because of the lower sampling rate. 60 ms is the highest value that can be used in the actual configuration. The movements become shaky with a higher value because the prediction of the needed acceleration is not accurate enough for such long periods of time. The variations in the dynamics are simply too high in this case. On the other hand, a lower value generates too much traffic to the robot and may cause problems if the generation of the next posture lasts longer than the StepWidth. In this case a new posture would need to be sent although the StreamAnimation would not have finished generating it. If this happens once , it can be compensated over the next steps by reducing the time between the postures until the delay is compensated. But if the generation causes delays too often, the animation cannot finish in time. The time that is needed to generate a posture of an animation depends on the complexity of the command similarly to the time that is needed to execute a posture command. In general, more time is needed to generate an animation posture than executing a posture command, but the time that is needed even for complex animations is much less then the critical 60 ms and not more than 30 ms for up to 32 joints. In Figure 7.8 the time that is needed to generate one posture during an animation depending on the number of used joints is plotted. It turned out that 40 ms achieves the best results.

7. SYSTEM DESCRIPTION

In special cases, like in the generation of lip movements it can be useful to reduce the value to not skip visemes. Here a `StepWidth` of 30 ms is used.

The last components that are needed to establish communication with the robot are the `JointGroups`. These are little modules that realize direct communication. They are able to interpret the translated PiRA-XML commands for a single joint, and can set the desired values directly in the robot. In some cases, it makes sense to have the option to use different controlling strategies for different parts of the robot. For example, the arm joints need to be controlled which requires fast bidirectional communication, and to control gaze a special module is used that communicates via YARP bottles. This is why dedicated components are used for each `JointGroup`. Each of these can realize its own communication strategy. Two strategies are used in the `CommandServer`: the `Bottle` and the `PolyDriver`. They are realized as abstract classes that can be implemented by the single `JointGroups`, and are generalized by an interface so that the `SetCommand` can use the same interface for all joints. Both are YARP components, and use the same means of communication with the robot, whereas the `PolyDriver` provides a more high-level interface.

A YARP `Bottle` is a container that holds the commands for the iCub. Since YARP is another middleware, the communication strategies are similar to XCF. The `JointGroups` connect itself to so called `Ports` that are provided by the software of the robot, which makes it possible to send `Bottles` to the joints to control them. This communication is unidirectional and cannot receive information about the joint such as its actual position, etc. To realize this, a second pipeline has to be used for the other direction to publish the data continuously similarly to the `XCF::Publisher`. The `PolyDriver` is therefore a better tool because it allows bidirectional communication which means that beside sending commands it is possible to request information as needed. For example, to compute the acceleration for the joints during an animation the actual position is needed.

The `JointGroups` that uses `Bottles` are used to realize gaze and facial expressions. Facial features like mouth and eyebrows on the iCub are displayed using LEDs and only support discrete values. For these there are no transitions between two postures and therefore no acceleration has to be computed to realize the trajectory. The `Bottles` are sufficient to realize the display of facial expressions. Gazing has to use `Bottles` because it is realized by using the `iKinGaze` module provided by the iCub community which only supports `Bottles` for communication. This module controls eye-neck coordination by itself, and is directly connected to the robot. All other `JointGroups` are realized by `PolyDriver` because they have to be controlled during the animation. These are groups for the head, torso, left arm, right arm, left leg and right leg.

7.7 Preprocess

The `Preprocess` is the last component of the `HardwareLayer` and has the task of interpreting the information collected by the sensors and providing it by inserting it into the `PerceptionMemory`. The actual system only uses the cameras that are located in the eyes of the iCub, and only these data are further evaluated for additional information. In principle, additional preprocessing of other sensors, i.e. touch sensors, can be integrated. To interpret the camera data, three modules are used: face detection, the saliency module and a face recognition system. All necessary camera data are provided by an `XCF::Publisher` which streams the camera image to each module that makes use of the data. After computing all high-level data, the single modules insert these into `PerceptionMemory` where they are accessible to the rest of the system over the high-level interface.

All three modules are realized using IceWing [91]. This is a configurable image pipeline that makes it possible to connect modules that have certain functions capable of building a complex image processing module. For example all modules used in the SLA use an import module to receive the camera stream and an export module that inserts the computed data into `PerceptionMemory`. Between these two modules, the feature extraction is done in a separate module.

To realize face detection, a detector is used based on the AdaBoost algorithm [150]. This module provides a list containing all faces within view of the robot. For each face, data are given for position, reliability and size. All of this is wrapped up in an XML structure as follows:

```
<CLUSTERS imageHeight="600" imageWidth="800" dbxml:id="1161" xmlns:dbxml=
                                    "http://www.sleepycat.com/2002/dbxml">
  <CLUSTER direction="-1" size="1">
    <HYPOTHESIS>
      <RATING>
        <RELIABILITY value="0.8"/>
      </RATING>
    </HYPOTHESIS>
    <REGION>
      <CENTER x="165" y="440"/>
      <RECTANGLE h="149" w="149" x="90.5" y="365.5"/>
```

7. SYSTEM DESCRIPTION

```
    </REGION>
  </CLUSTER>
</CLUSTERS>
```

This structure is inserted in `PerceptionMemory` each time a new frame is evaluated. The face detection module only detects faces that face the robot directly. This was done intentionally so that the robot can only react to users that attend the robot.

The saliency module used is an implementation of the saliency module by Itti and Koch [78]. It computes a feature map of interesting regions within the image. Three different criteria are used to define what is interesting. First, the colorful regions have a high value for interest. Second, regions with a high intensity also generate a higher interest than regions with lower intensity. Lastly, the orientation of structures in the image are rated. Those regions that have a higher interest have an orientation that differs from the orientation of the surrounding regions. For all criteria, feature maps are generated that are combined into one saliency map that considers all feature maps. The salience map contains a salience value for each region in the image. The region having the highest value is the region which attracts the most interest. The saliency map in the used implementation is represented as an XML structure containing a list of points that are sorted by their saliency value beginning with the highest value. This XML structure is inserted into `PerceptionMemory` after each computation step. A sample structure may look like the following:

```
<saliencies width="50" height="37" dbxml:id="61" xmlns:dbxml="http://www.sleepycat.com/2002/dbxml">
  <TIMESTAMP>
    <INSERTED value="2039825843"/>
    <UPDATED value="2039825843"/>
  </TIMESTAMP>
  <region cx="5" cy="28" saliency="64033" colorV="4056" intensityV="3872" orientationV="13230"/>
  <region cx="5" cy="29" saliency="55682" colorV="4732" intensityV="8712" orientationV="7938"/>
  <region cx="9" cy="27" saliency="53746" colorV="3380" intensityV="7260" orientationV="2646"/>
  <region cx="2" cy="27" saliency="52971" colorV="2704" intensityV="3388" orientationV="3969"/>
  <region cx="0" cy="27" saliency="37857" colorV="2704" intensityV="6292" orientationV="9261"/>
  <region cx="39" cy="24" saliency="34803" colorV="23660" intensityV="6292" orientationV="4851"/>
  .
  .
  .
</saliencies>
```

The last component of the `Preprocess` is face recognition. Although this module is not used within this thesis it has been used in combination with other `FunctionalLayers`. One example is a system that can detect interaction episodes and automatically learns the faces of important persons to build up links between persons and past interaction episodes [58]. To realize the person recognition the `caiwicat face recognition` `system` was used that was invented by Lang et al. [64]. This system uses an Active Appearance Model approach to classifying previously learned faces and provides the function to learn new faces and relearn already learned faces during runtime.

7.8 Social Layer

The `SocialLayer` contains all the social competences and generates all DBCs that are used by the SLA. Most of its functions are generated automatically i.e. blinking and idle DBCs. Some of these components can be influenced by sensor data (e.g.: gazing behavior) or the internal state provided by the `LongTermMemory`. Other components can be accessed directly by using the high-level interface (e.g.: speech). The generated output of this layer is sent to the `Arbiter`. The `SocialLayer` generates its behavior continuously when activated. The `Arbiter` decides if the behavior can be executed on the robot, or suppresses it if commands with a higher priority are blocking necessary joints. This ensures that the robot displays the social behavior only when no conflicting functional behavior is being carried out. This layer consists of four main components: the blinking DBC, the gaze DBC, the idle DBC and the lip DBC. In addition a small `EmotionServer` is integrated that displays facial expressions depending on the emotional state of the robot. Furthermore, the emotional state is influenced by whether the robot detects faces. The robot becomes "happy" if someone is looking at it, and "sad" if no human can be detected. This behavior is not evaluated here, and therefore, was not used in the studies.

In the following section, the generation of the DBCs will be explained briefly. The theoretical background and patterns used are described in chapter 6. The generation of gaze, blinking and idle DBCs is very simple because they depend largely on producing predefined movements with random timing.

The gaze DBC needs two data streams to generate its behavior. The position of detected faces, and the saliency data are both available in `PerceptionMemory`. The gaze module subscribes to both streams and gets informed if new data are available. It is always aware of where humans are and where interesting regions are located. As described before, the gaze pattern shifts from time to time between human faces and salient regions to reduce discomfort. The time elapsed between two shifts is based on a random process, and is defined by three parameters. The first parameter

7. SYSTEM DESCRIPTION

defines the mean duration of mutual gaze, and the second the mean duration of gazing elsewhere. The third parameter defines the variation of the duration as a percentage of the given duration. For example a duration of 4 seconds with a variation of 25% results in possible durations ranging from 3 to 5 seconds which are randomly chosen after each gaze shift. To perform a gaze shift that terminates the mutual gaze, the robot needs to be made to look in a different direction. The problem is that faces often are prominent regions for the saliency module within the field of view. To prevent the robot from looking continuously at the human face, the region of the face is occluded, and the next most salient point is chosen that is not the interaction partner. It might happen that face detection does not detect a face because there is a malfunction or simply because there is no face to detect. In these cases, the gazing behavior continues looking at different salient regions as an idle movement until a new face is detected. If neither face data not saliency data are present, the robot keeps looking in the last direction.

The implementation of idle movement is based on a set of predesigned movements randomly chosen to be carried out by the robot. As mentioned before, these are breathing, looking around and a stretching motion. After each breathing animation, one of the two idle motions is carried out with a probability of 15%. All animations that are used to move the robot are stored in the BMD, and are robot independent.

The blinking DBC performs blinks from time to time to. This behavior uses the same randomization pattern as the gazing behavior. The mean time between both blinks and the variation can be set using two parameters. A third parameter defines how long the eyes are closed. This duration has no variation.

The exact parameters used and a detailed description of the movements is described in chapter 6

The fourth is the lip DBC, and an example of the high-level interface. This module provides an XCF Server to send speech commands to the robot. As input, only a plain text is needed to perform speech synchronous lip movements. This text is used to generate a phoneme sequence via MARY. This sequence is, on the one hand used, to generate an audio file that contains the speech that has to be be played back. On the other hand, a viseme sequence is generated in parallel that defines the timing of the lip movements. The final translation of visemes into robot commands is done afterwards in the HAL which requests the corresponding joint commands from the BMD. A more detailed description of the theoretical generation of lip movements is done in section 6.5. To play the sound and the movements in synchrony, a synchronization step is required. After generating both streams, the audio device triggers when it is ready to play the sound file. This signal is used to synchronize the beginning of the streams. As soon as the play back starts the generation of lip movements starts, too. For one sentence or a few short phrases no synchronization is necessary during the stream. For longer speech, it is advisable to split it into several commands with shorter sentences. Nevertheless the drift of the two streams is not reasonable. A text with 350 words (2 minutes of speech) has an average shift of less than 500 ms. Another feature of this high-level interface is that it can integrate any PiRA-XML command into speech to add movements. The movements start as soon as the speech begins. This makes it possible to execute speech with synchronous lip movements and gestures with a single command. Putting a sequence of these commands together can be used to generate a whole monologue for the robot. In the Appendix, Figure D.3 shows a sequence diagram for a typical speech command with integrated PiRA motion. The structure of the command to be sent to the robot follows a similar structure as the PiRA-XML files and looks like the following:

```
<SpeechData language="eng" lips="true">
        My name is iCub.
        <Animation>
                <PiRAxml robot="iCub" robotType="humanoid" type="animation" duration="3000">
                        <Joint jointType="SimpleJoint" name="rightarm_shoulder_pitch">
                                <Anchor time="0" value="7" />
                                <Anchor time="719" value="-31"/>
                                <Anchor time="2074" value="-31"/>
                                <Anchor time="3000" value="7"/>
                        </Joint>
                </PiRAxml>
        </Animation>
</SpeechData>
```

The command is surrounded by a tag named `SpeechData` which has two optional attributes. The attribute `language` defines the language which will be used to interpret the given sentence
(ger=German; eng=English(default)). The second attribute, `lips`, defines if the lips will be used during speech or not (true=withLips(default); false=noLips). Within the tag `SpeechData`, the sentence is specified. In addition, the optional tag `Animation` can be used to integrate any valid PiRA-XML command.

At the moment the lips are the only implemented high-level interface. More will follow in the future. As already mentioned, it would be useful to gen-

7. SYSTEM DESCRIPTION

erate arm movements by setting the 3D coordinates in robot space. A disadvantage would be that these movements often are not natural when only kinematic algorithms are used. As an additional high-level feature, it would be reasonable to enrich the functional arm movement automatically with a more natural movement as a background cue as is done manually with the editor at the moment.

7.9 Exchangeability

In this section some tests are presented that were done to test whether the system fulfills the exchangeability criteria. The first proof was the fact that the SLA was developed on different robots. The first versions of this architecture were developed for Barthoc Jr. and the Sony AIBO. After around one year, the Flobi head was available, and the first implementations on the iCub were made. In all cases, only components of the `HardwareAbstractionLayer` had to be changed. In the main phase of this research, it was decided to use only the iCub as a research platform although drivers for other robots are available. The dynamic development of the SLA on various robots shows the high exchangeability of this system. Another test that was carried out was to connect a Smart Phone to the SLA as a functional layer. After only 30 minutes of programming, the Smart Phone was able to function as a remote controller for the robot. This remote control app contains several buttons that can trigger various animations such as nodding, and can have the robot talk with synchronized lip movements.

The current plan is to provide additional interfaces to help integrate various types of middleware to the SLA e.g. Ros [124]

7.10 PiRA-Editor

The PiRA-Editor is a Graphical User Interface (GUI) used to easy edit PiRA-XML files. This tool can display animations in an easy understandable way, and allows changes to be made without having to rewrite the code directly. Furthermore it is possible to save, load and export animations and to test them directly on the robot while editing. In this section the functions of the PiRA-Editor are presented and a usability study is described that compares the usability of the PiRA-Editor with that of a similar editor called *Choregraphe*. In addition other editors that have similar functions will also be described.

7.10.1 Program description

The idea behind designing this tool was to make it possible to edit animations without having to write the PiRA-XML file manually. It is possible to add different joints as needed, and to configure the desired position of each joint over the entire time span of the animation. This is done by placing **Anchors** onto a timeline which represent the angle of a joint at a certain point in time the same as the **Anchors** used in PiRA-XML files. This concept is part of the goal of providing a high usability to the developer, which is one of the criteria for social robots. The PiRA-Editor was created to achieve this goal. The GUI for this tool is shown in Figure 7.9. On the left side of the program all necessary joints can be chosen and can then be configured on the right side of the screen. The main purpose of this tool is to export the configured animation as a PiRA-XML file that can be used within the SLA. This makes it very easy to create robot movements for modules that are part of the **FunctionalLayer**. In addition, it is possible to test the actual animation directly on the robot by clicking the "Run Animation"-button to improve the animation before exporting the final PiRA-XML.

To realize compatibility in different robots the PiRA-Editor consists of a front end which provides the GUI, and a driver which provides the necessary data for each robot, such as available joints, and establishes communication with the robot for the "Run Animation" function. When using the SLA, this driver is always the same because the robot dependent code is located in the **HAL**. In theory, the PiRA-Editor is compatible with different robots that can be controlled directly. For example the Sony AIBO can be controlled with this tool as well.

To better work with the editor it is possible to save and load animations in a file format similar to the PiRA-XML file, although it contains additional information necessary for the editor. When exporting the animation a pure PiRA-XML file is created that can be used within the SLA directly. In addition, a help function can be used should some functions remain unclear.

7. SYSTEM DESCRIPTION

7.10.1.1 Graphical User Interface

In this section the GUI of the PiRA-Editor is described in detail. In Figure 7.9 the GUI is shown with a sample joint configuration. When creating a new animation the `JointConfiguration` window is empty because no joints have been added so far. First, it is advisable to adjust the length of the animation as needed, which can also be done later during editing. It should be noted, however, that future adjustments may be limited by constraints depending on the anchors that have already been configured. For example, the animation time cannot be shortened to a length shorter than 2400 ms if an `Anchor` has already been defined at a time later than 2400 ms.

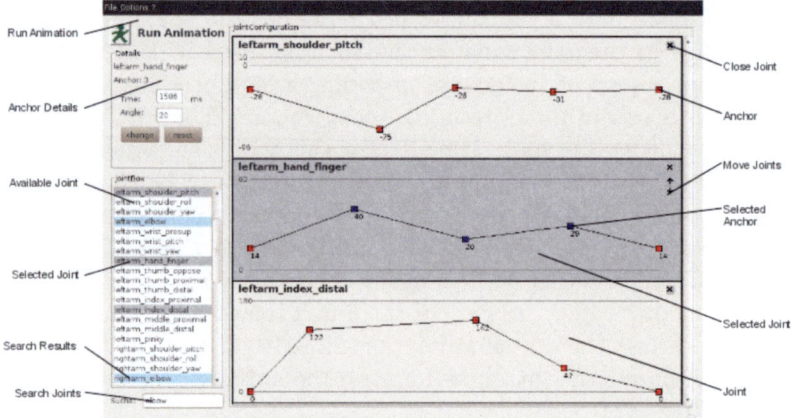

Figure 7.9: Graphical User Interface of the PiRA-Editor. This Figure shows a sample joint configuration for an animation with a length of 3000 ms.

The `JointBox` provides an overview of all joints that are available. By double clicking on one of the joints, it can be added to the `JointConfiguration`, becoming active for the current animation. This means that this joint blocks resources when the command passes the arbiter, and has to define joint angles for the span of the animation. Because of this, default values are loaded when adding the joint the first time to provide a valid joint configuration. These can be modify to realize the final animation. The default values are those which used to define the social posture (see section 6.3). If a joint is not needed anymore, a second double click removes the joint after answering a confirmation dialogue ("Do you really want to close this joint"). All joints that are used within the animation are highlighted with a gray background. When pressing `Ctrl-F`, a search window appears beneath the `JointBox` which allows the user to enter a keyword. After pressing the `Enter`-key all joints that contain the keyword are highlighted in turquoise.

After inserting all necessary joints, they are listed in the `JointConfiguration` window on the right side of the GUI. In this list, the configuration of the `Anchors` takes place. Each segment represents one joint and has a time line starting at 0 ms on the left and ends at the duration of the animation (3000 ms in this example) on the right. All joints use the same time line, and therefore `Anchors` that are placed exactly above each other have the same timing. The other dimension represents the possible angles for each joint starting with the minimum value at the bottom and the maximum value at the top of the segment. In some cases, the joints do not have a continuous range of values, as in, for example, lip postures. Here, a range on a time line can be configured to represent the duration the state should be sustained. Figure 7.10 shows the configuration of such a discrete joint. When the joint is selected, three buttons appear in the top right corner (Otherwise only one button). The button with the "x" closes the joint. The buttons with the arrows move the joints one position upwards or downwards.

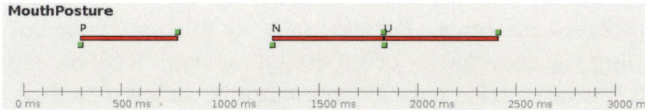

Figure 7.10: Example for a discrete joint. The visemes "P", "N" and "U" are configured to be displayed on the robot

When configuring each joint, different possibilities are available to modify the `Anchors`. By right clicking on any position on the time line, a new `Anchor` is created at that position. The position defines the time and the value that the joint should have at this point. In the example the second `Anchor` for the "leftarm_hand_finger" defines that this joint should have the value 40 at the given time. By using the third mouse button, an `Anchor` can be deleted. By performing a left click it can be selected. To select multiple `Anchors` they can be added by Ctrl-clicking on them. To manip-

7. SYSTEM DESCRIPTION

ulate the position of an `Anchor`, it can be dragged with the mouse. It is also possible to select multiple `Anchors` at the same time to move them in parallel as long as none of them is moved into an invalid area. In this case the `Anchors` can not be moved further.

Another important component in the GUI is the `Details` window, located in the top left corner. This window shows detailed information about the selected `Anchor` and allows new values to be entered. In addition it is possible to enter new `Anchors` for the selected joint. The last element is the "Run Animation"-button above the `Details` window. This element was explained above and executes the actual configured animation on the robot. After finishing the animation, it can be saved or exported as PiRA-XML by using the `File`-Menue.

7.10.1.2 Similar Editors

Beside the PiRA-Editor, other editors exist that are capable of doing similar tasks. In first place, Choregraphe [123] invented by Aldebaran Robotics, should be mentioned. This program is a complete graphical environment for the Nao robot, and provides many functions to design the behavior of the robot. This is done by connecting different high-level behaviors to generate a complex behavior using a GUI. These behaviors are mostly animations that can also contain speech to move the robot. Furthermore, they can be connected in such a way that they are executed one after the other, allowing for the creation of loops. To generate the behaviors Choregraphe provides different methods. For example, by moving the robot manually and recording the movements or by designing them with an editor similar to the PiRA-Editor. This part of Choregraphe, called *Timeline*, is the key part in this section.

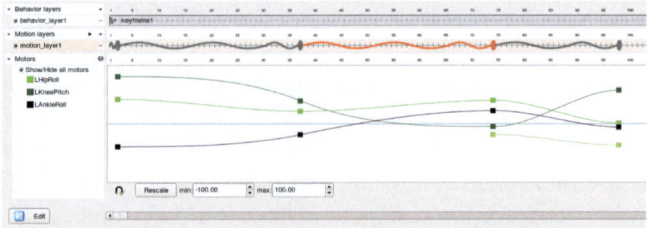

Figure 7.11: Timeline window from Choregraphe

The Timeline allows editing of previously created animations and the generation of new ones. Figure 7.11 shows Timeline with some joints already defined. Timeline uses a similar representation of the joints as PiRA-Editor. The difference is that here all joints are displayed in one graph and with colors used to identify them. Choregraphe, in addition has the possibility of adding speech, and structuring different movements into different layers. It is possible to define the arm and the leg movements separately. To edit the different joint movements Choregraphe uses an interesting approach. It is possible to directly click onto the joint group in a simulator for the Nao, and to set joint positions in an additional window that displays the joint group (e.g. a leg) to be configured. All changes are directly visible in the simulation so that the posture can be checked directly while configuring it. To sum up, Choregraphe is a powerful tool but in some cases confusing for the novice as shown in a usability study comparing PiRA-Editor and Timeline from Choregraphe in section 7.10.2.

Another software that is capable of designing robot motions is the SDR Motion Creating System [88] which was created to design the dance performances of the SDR-4X. This animation editor consists of two parts. The configurator for the upper part is used to design complete movements for all joints above the hip. The lower part of the robot is controlled by a separate part of the program. This part provides stable lower body motion to prevent it from falling, and is capable of creating walking motion that is synchronized with music. Both the SDR Motion Creating System and Choregraphe have the disadvantage that they are only compatible with one robot or one robot series.

Saerbeck et al. [135] also designed an editor for robot animation by using an interesting approach. This editor uses key frames to define the animation, but in contrast to many other editors, it does not simply interpolate between key frames. Instead they invented an algorithm to generate a natural interpolation between key frames to make the animation more natural.

7.10.2 Usability Study

The goal of this usability study was to test if the GUI design of the PiRA editor fulfilled the third criteria of social robots which says that the software

7. SYSTEM DESCRIPTION

must have good usability for the developers. To achieve this, participants were asked to solve some tasks with the editor. We evaluated how long this took and what problems were reported. None of the participants had ever used the PiRA-Editor before, and most had no experience with controlling robots using similar software. This was done to test if the user interface was intuitive and easy to use even for inexperienced users.

In addition, the PiRA-Editor was compared with a commercial product. It was not expected that the editor would be better than the commercial product, since much more time and man-power is typically invested in commercial software. Regardless, the idea was that the comparison would provide a good estimate of the PiRA-Editor's performance. For this test Choregraphe was used as the commercial product. To keep it comparable only Timeline from Choregraphe was compared with the PiRA-Editor as both have similar functions.

7.10.2.1 Setup and Measures

In this usability study, two conditions were used, each defined as use of either the PiRA-Editor, or Choregraphe. For both tools participants had to solve very similar problems. The only reason that the tasks vary a little was because of different usability concepts in each program. The differences were kept to a minimum to facilitate comparison. The same laptop was used for both conditions, including the same mouse. The software was prepared by the experimenter so that all participants could directly start solving the tasks. In addition a simulation of the robot was available in both conditions to check the animations. The participants were told what kind of program they were using, and that they could control a robot with it. In addition, they received instructions to solve the tasks as quickly as possible.

The tasks were to familiarize themselves with the program, and to create animations for the robot. All tasks were plotted on an exercise sheet so that participants could work on their own. Only when problems occurred were they allowed to contact the experimenter. The tasks were divided into three parts: first steps, imitating a movement, creating an animation. The "First steps" section consisted of six subtasks to help participants get used to the program. In some cases, useful tips were provided to help with certain functions as necessary. To solve these tasks, the participants had to insert the joint for the right elbow into the animation and add 5 anchors to generate a movement. After that the left elbow was to inserted so that it performed the same movement as the right. This was followed by another subtask to edit an anchor, and set a given value and time. Here the participants were asked to make use of the edit function allowing them to enter values from the keyboard without having to drag the anchor with the mouse. For the PiRA-Editor, the additional task was given to select multiple anchors and move them to the highest possible value, a function not available in Choregraphe.

In the "Imitating movement" section participants were asked to load a predefined waving movement and edit so that the robot performed the motion with the other hand. The last section asked the participants to create a new movement on their own. In this case was a special bow movement called "Kratzfuß"-bow in German. What is special about this movement is that beside the normal bow, one hand is placed in front of the stomach and the other is extended to the side. A picture is provided on the exercise sheet to clarify this movement. During all tasks, the participants are instructed to make use of the simulator to check their animations. The whole study was carried out in German. The original exercise sheets can be found in the Appendix A.3. For the PiRA-Editor the participants got a draft with pictures of the iCub where all joints are labeled to identify them in the editor. For Choregraphe such a draft was not necessary because the joints can be identified by clicking on it in the simulator.

The measure used was the time that was needed to solve the task. Only sections two and three were evaluated because the training task in section one differed too much to be comparable. Especially because of the additional task for the PiRA-Editor.

After completing the tasks, participants were asked to fill out two questionnaires. The first consisted of several questions about the software itself. The questions were:

- How easy is the handling of the Editor? (3 point scale)

- Is the arrangement of the single components intuitive? (3 point scale)

- Is the naming of all components and menu items appropriate? (5 point scale)

The highest score indicated total agreement. In addition, the participants had the option of giving comments to the following three questions:

7. SYSTEM DESCRIPTION

- What did you think is positive referring to the handling of the program?

- What did you think is negative referring to the handling of the program?

- Miscellaneous comments.

The second questionnaire consisted of several questions to personal details of the participant such as age, computer experience and so on.

7.10.2.2 Results and Discussion

In the usability study 14 participants (2 female, 12 male) with an age ranging from 19 to 30 ($M = 23.86$, $SD = 3.939$) participated. All participants had high computer experience with a mean value of 4.71 ($SD = .469$) on a 5 point scale. This was a between-subject design, with 7 participants testing the PiRA-Editor and 7 Choregraphe.

To compare the time (in minutes) the participants needed to solve section two and three independent-sample t-Tests were used to test for significant differences. For the **second section** there occurred a significant difference between the PiRA condition ($M = 10.43$, $SD = 1.397$) and the Choregraphe condition ($M = 15.57$, $SD = 4.962$), $t(12) = -2.64, p = .022, d = 1.41$. For the **third section** there was no significant difference between PiRA-Editor ($M = 23.14$, $SD = 11.393$) and Choregraphe ($M = 18.71$, $SD = 7.847$), $t(12) = -0.847, p = .414, d = 0.453$.

To evaluate the questionnaires a Mann-Whitney Test was carried out to test for significant differences. There were no significant differences in the **handling** of the PiRA-Editor ($Mdn = 2$) and Choregraphe ($Mdn = 2$), $U = 17, p = .298, r = 0.278$. The arrangement of different components was rated as more **intuitive** in the PiRA-Editor ($Mdn = 2$) than in Choregraphe ($Mdn = 2$), $U = 8, p = .018, r = 0.63$. Lastly there are no significant differences in the **appropriateness of the labeling** between the PiRA-Editor ($Mdn = 2$) and Choregraphe ($Mdn = 2$), $U = 13.3, p = .114, r = 0.422$.

These results show that there was not much difference between the two editors, indicating that the PiRA-Editor is rated equal to a commercial product. In some cases the PiRA-Editor is rated better than Timeline from Choregraphe. Participants stated that the arrangement of the components are more intuitive in PiRA, and the second section was solved faster than in Choregraphe. The reason for this might be that the PiRA-Editor has fewer functions compared with Choregraphe, and is therefore easier to use. Choregraphe is a professional and complex tool, and may often not be intuitive to novice users. Nevertheless this results show that the participants rated the PiRA-Editor as intuitive and easy to use, and it shows that it is capable of competing with Timeline.

In addition to the quantitative data, participants were asked to give comments on both programs. In general, these were largely positive for the PiRA-Editor. Some participants said that it was fun to use the editor and most said that it was easy to use and flexible. In addition, the `Details` window was seen as a comfortable alternative to the dragging of anchors. On the other hand other minor usability issues were mentioned that could be improved. For example, it was suggested to include the option to hide joints without closing them because scrolling through a longer list of joints becomes uncomfortable. In addition, one participant would prefer to be able to move the joints with the mouse instead of the buttons. In general no serious usability problems were reported.

The comments for Choregraphe were similar to the other editor. Many participants also said that the tool was easy to use though it took some time to get used to. Another positive aspect was the simulation that allowed changes to be seen immediately. One participant liked the hierarchical representation of the joints saying it made it easier to navigate through the joint list. In some cases it seemed that the participants had problems using the tool and its functions correctly. Three participants seemed to have major problems while using this editor and gave comments like "unacceptable and buggy" and "unacceptable handling". In general, slightly larger problems with usability were reported for Choregraphe as compared with the PiRA-Editor. In particular, the representation of the trajectories of the joints using only one diagram was rated as confusing, further complicated by missing labels on the anchors and the time line. In addition the editing of the anchors was described as unintuitive and random by some users.

In summary, it can be said that both tools are rated as very good although some participants had problems with Choregraphe. The general usability of the PiRA-Editor was rated as slightly better in the comments and only participants that used the PiRA-Editor said that they had fun using it.

7. SYSTEM DESCRIPTION

This usability study shows that the usability of the PiRA-Editor is already very good and revealed some ideas about what could be improved. For example, the hierarchical representation of the joints, like those in Choregraphe, would be an interesting improvement. Most of the participants had fun while using it and no one had major problems, resulting in very positive ratings of the handling and intuitiveness of the editor. The positive comments and constructive criticism further reinforce the positive rating. Direct comparison with a commercial product showed that the PiRA-Editor is a competitive product. Its usability and handling was rated as equal or in some cases better than Choregraphe. All this shows that the PiRA-Editor fulfills the third criteria for social robots and is a very useful tool for designing natural movements.

Chapter 8

Studies

In the previous chapters, the creation of Dynamic Background Cues (DBCs) was presented, and it was discussed how these cues could influence Human-Robot Interaction (HRI) when robots appear more social by displaying DBCs. This theoretical discussion led to the construction of DBCs with the intention of influencing the human so that the robot appeared more familiar and lively without effecting its functional behavior. Furthermore, it was hypothesized that the social behavior of robots influenced the arousal of the human participant when interacting with the robot. During several interaction and online studies, a set of DBCs was created to test the hypotheses. In this chapter, two studies are presented that combine the previous work by examining all remaining hypotheses.

The first study is another online study that test if the combination of several DBCs is able to generate robotic character traits. This investigation evaluates the last part of the second Hypothesis, and tests if it is possible to create the desired robotic character for the interaction study.

The second study tests the third hypothesis. The users participated in a interaction study in which the robot iCub gave them instructions to complete some simple tasks. During the experiment biophysical measures were collected to see if the arousal state of the participants varied depending on the amount of social behavior exhibited by the robot. To generate the two conditions, the previously optimized DBCs were used. In addition, task performance was evaluated to test if DBCs influenced the time needed to solve the tasks.

8. STUDIES

8.1 Whole Body Study

In this section, another online study is presented to evaluate the effects of all presented DBCs in combination on how robots are perceived. The previous online studies tested single DBCs as part of the first argument in the second hypothesis. This hypothesis states:

Hypothesis 2. *The use of parameterized DBCs makes it possible to create certain robotic character traits that are perceivable by human users both when using single and multiple cues in combination.*

The following study aims to test the second part which argues that DBCs in combination can have the same effects on people as separate DBCs. Here, care must be taken that single cues have the same effects on the human participants, and do not disturb the overall impression. If two cues present two different character traits (like friendly and harsh), this may cause expectation mismatches because these cues do not fit together. To prevent this, the research from the previous online studies (see section 6.4.1 and section 6.5.1) was used to generate a friendly personality. In addition, this study provides another opportunity to test the first hypothesis which says that DBCs can help a robot appear more friendly and lively.

8.1.1 Setup

The setup used for this study was the same as for the other two online studies (see chapter 6.4.1). All participants were asked to view two movies of the iCub (within-subject study design). One had DBCs turned on, and the other had them turned off. After each video, a short questionnaire containing 26 items was filled out before watching the next movie. The complete study was designed such that every participant was able to participate on their own using their own computer. To achieve this, a PHP-Script was written to display videos and questionnaires whose results were saved in a mySQL-Database for further evaluation. The URL of this study was distributed via several social networks and email to acquire a broad variety of people. On the first page of the website, the study was introduced and the participants were told what to do. They were also instructed to conduct the study alone and to turn on the sound. The original German introduction was very similar to the introduction given in the previous online studies (see Appendix Figure A.4).

After confirming the comprehension of the introduction, the first video was shown which was randomly chosen from the two conditions. The videos were viewed twice attentively (see Appendix Figure A.5). On the next page, a questionnaire about the video appeared. The video became unavailable as soon as the questionnaire appeared so that responses depended strictly on its personal feeling and memory. The questionnaires contained 26 items, and will be described in the following section. After filling out the questions, the next video was shown, followed by the second questionnaire that belonged to the other condition. In the end of the session all participants were asked to fill out some personal data including age, gender, native language, job, highest degree completed, computer experience, robot experience and if they had already participated in a study with iCub. In addition the Big5 scale was used as described in chapter 5.

In both conditions the robot iCub was shown performing a short monolog. The angle of the camera was chosen such that it showed the front of the robot from the head to the knees in front of a blue curtain. The difference between the conditions was the social behavior of the robot. In the non-social condition the robot only performed functional behavior and in the social condition this functional behavior was enriched with several DBCs. The functional behavior of the non-social condition was as follows. First the robot said: "You have to go this way" and pointed in a direction to the right of the robot. This sentence was chosen because it is very neutral, does not convey any emotion, and is of an informative nature suited to the other functional signs. In addition the lips were not moved during speech. The pointing gesture was reduced to its functional core, as well. To perform the pointing behavior the robot only used the two shoulder joints to move the hand in the desired direction, and turned the head into the same direction using only the jaw joint. The movement is shown in Appendix Figure C.4 on the right. After the pointing gesture, the robot looked around by shifting its gaze from right to left. In the non-social condition this was done by only using one neck joint to turn the head from one side to the other. The eyes did not perform any movement relative to the head during the entire procedure.

For the social condition the same functional movements were used as for the non-social, although these movements were enriched by using the DBCs that were described in chapter 6. This means that in both conditions, the robot provided the same information as the functional signs did not changed. First, the lips were animated while speaking. Second, while gaze

8. STUDIES

shifts are carried out, the robot used natural neck-eye coordination. In addition, the robots showed simulated breathing at all times. As can be seen in the Appendix Figure C.4, on the left side, the pointing gesture was changed so that it used a natural movement style by bending the arm a little before lifting and stretching it at the end of the pointing gesture. The last difference was that during the looking around animation, the whole body of the robot was used. The robot leaned forward and turned the entire upper body in the direction of its gaze, and moved the arms accordingly so that they hanged down with physical correctness. This made it appear more natural without adding additional information. In Appendix Figure C.2 this animation is shown.

8.1.2 Measurements

The questionnaires that were filled out during this study consisted of 26 items, and were completed once for each condition. The questions were mainly items taken from the other two online studies, or inspired by the questionnaires given by Parise et al. [119] and Warner et al. [155]. They were divided into three groups. In the first group the character traits of the robot were considered. The second group consisted of items concerning the movements of the robot directly. Questions that rated how the participant perceived the relationship between the robot and herself were part of the third group.

Questions concerning the character traits of the robot consisted of 14 items with opposing traits on each side (e.g.:relaxed(1)/nervous(6)). Each item used a 6 point scale to force participants to choose one or the other side. Five of the 14 items were taken from previous online studies to make all of them comparable. First, the three VAD items were used (positive/negative, calm/aroused and submissive/dominant) to rate the robot. Second, the two items disinterested/attentive and positive/negative were also adopted. The additional nine items were included to allow a more detailed evaluation of how the robot was perceived. For example some items were inserted to rate the perceived intelligence of the robot (e.g.: unintelligent/intelligent, incompetent/competent, or unreliable/reliable)
In the second group, the items were sorted to directly correspond to the perception of the movements. Here three items were used that were rated on a 6 point scale. The questions to be answered were: "The movements of the iCub..." ...were stiff/...were elegant, ...were pleasant/ ...were unpleas-

ant and ...were not disturbing/...were disturbing. Here, the first was rated with a score of 1 the second with 6. The reason to insert these items was to figure out how the movements in general were perceived by the human participants.

The items that were used in the third group were mainly inspired by the studies carried out by Eyssel et all (e.g.: [52]). Here nine items were used that were rated on a 6 point scale ranging from strongly disagree(1) to strongly agree(6). These items focus on the perceived relationship between the robot and the participant. Some sample items were "How close do you feel to the robot? ", "How human-like did the robot appear to you? " or "Do you trust the information given by the robot? ". These questions were used to gain more insights onto the relationship between the human and the robot. To the item "How human-like did the robot appear to you? " the participants had the option to leave comments that were used for the qualitative evaluation.

The original German questionnaire can be found in the Appendix A.4.

8.1.3 Results

The results for this study are as follows. 34 Participants (10 female, 22 male, 2 not specified) took part in this study. The age of all participants ranged from 16 to 75 ($M = 38.88$, $SD = 15.79$). The participants rated their own computer and robot experience on a scale ranging from 1 (none) to 5 (very much). They had a mean computer experience of 2.94 ($SD = 2.03$) and a mean robot experience of 2.32 ($SD = 1.22$). Because of the study design all participants rated both conditions.

For the evaluation of the questionnaires a Wilcoxon Signed Rank Test was used to test for significant differences between the conditions. A detailed list of results can be found in Appendix B.5. For the item **unattractive/attractive** the social ($Mdn = 4$) condition was rated as more attractive than the non-social ($Mdn = 4$) condition, $Z = 2.62, p = .009, r = 0.45$. In addition the **emotional/callously** item was rated as more emotional in the social ($Mdn = 3$) as in the non-social ($Mdn = 5$) condition, $Z = 4.1, p = < .001, r = 0.7$. The item **unsocial/social** was seen as more social in the social ($Mdn = 5$) than in the non-social ($Mdn = 4$) condition, $Z = 3.01, p = .003, r = 0.52$. There were no significant differences for the item **dense/intelligent** between the non-social ($Mdn = 4$) and the social ($Mdn = 4$) condition, $Z = 0.85, p = .391, r = 0.15$.

8. STUDIES

The same was true for **relaxed/nervous** where no significant differences could be found between non-social ($Mdn = 3$) and social ($Mdn = 2$), $Z = 1.37, p = .170, r = 0.24$. For **incompetent/competent** the social ($Mdn = 5$) condition was rated as more competent than the non-social ($Mdn = 4$) condition, $Z = 2.87, p = .004, r = 0.49$. There were no significant differences for the item **calm/aroused** between social ($Mdn = 2$) and non-social ($Mdn = 2$) condition, $Z = 1.81, p = .07, r = 0.31$. For the social ($Mdn = 5$) condition the robot was rated as more active than for the non-social ($Mdn = 4$) condition for the item **passive/active**, $Z = 3.740, p =< .001, r = 0.64$. There was no significant difference for **submissive/dominant** between the social ($Mdn = 4$) and the non-social ($Mdn = 3$) condition, $Z = 0.32, p = .747, r = 0.06$. Furthermore in the social ($Mdn = 3$) condition the robot is seen as more positive than in the non-social ($Mdn = 3$) condition for item **positive/negative**, $Z = 2.76, p = .006, r = 0.47$. For item **disinterested/attentive** the social ($Mdn = 5$) was rated as more attentive than the non-social ($Mdn = 4$) condition, $Z = 2.82, p = .005, r = 0.48$. For the item **unreliable/reliable** there were no significant differences between the social ($Mdn = 5$) and the non-social ($Mdn = 4$) condition, $Z = 1.42, p = .156, r = 0.24$. In the item **natural/artificial** the robot was rated as more natural in the social ($Mdn = 3$) than in the non-social ($Mdn = 5$) condition, $Z = 4.25, p =< .001, r = 0.73$. The last item of this group (**unlikeable/likable**) was rated as more likable in the social ($Mdn = 5$) than in the non-social ($Mdn = 4$) condition, $Z = 2.65, p = .008, r = 0.45$.

For the three item of the second group, concerning the movements of the robot, the following results were gathered. The movements were seen as more **elegant** in the social ($Mdn = 4.5$) than in the non-social ($Mdn = 2$) condition, $Z = 4.06, p =< .001, r = 0.7$. Furthermore they were rated as more **pleasant** in the social ($Mdn = 2$) than in the non-social ($Mdn = 4$) condition, $Z = 3.46, p = .001, r = 0.59$. For the item **disturbing/not disturbing** the robot was seen as less disturbing in the the social ($Mdn = 2$) than in the non-social ($Mdn = 3$) condition, $Z = 2.66, p = .008, r = 0.46$.

Within the last group of questions the participants rated that they thought **the robot could become a friend** significantly more often in the social ($Mdn = 3$) than in the non-social ($Mdn = 2$) condition, $Z = 3.27, p = .001, r = 0.56$. Furthermore they **did feel closer to the robot** in the social ($Mdn = 3$) than in the non-social ($Mdn = 2$) condition, $Z = 3, p = .003, r = 0.51$. In the social ($Mdn = 4.5$) condition the participants **liked** the robot more than in the non-social ($Mdn = 4$) condition,

$Z = 2.77, p = .006, r = 0.48$. There was no significant difference for the item **Are you able to estimate the behavior of the robot?** between the social ($Mdn = 4$) and the non-social ($Mdn = 3$) condition, $Z = 2.31, p = .21, r = 0.40$. For the item **Would you like to interact with the real robot?** there was no difference between social ($Mdn = 5$) and non-social ($Mdn = 4$), $Z = 1.43, p = .154, r = 0.24$, but in both cases the score was relative high. The participants rated the robot in the social ($Mdn = 4$) as more **vivid** than in the non-social ($Mdn = 2$) condition, $Z = 4.29, p = < .001, r = 0.74$. In addition they rated the robot as **more human-like** in the social ($Mdn = 4$) than in the non-social ($Mdn = 3$) condition, $Z = 4.14, p = < .001, r = 0.71$. On the question of whether they **trusted the information given by the robot**, participants in the social ($Mdn = 5$) condition agreed more to this statement than in the non-social ($Mdn = 4$) condition, $Z = 2.01, p = .045, r = 0.34$. Lastly, participants thoughts more that the robots was able to **act on its own** in the social ($Mdn = 4$) than in the non-social ($Mdn = 2.5$) condition, $Z = 2.67, p = .007, r = 0.46$.

Beside testing for significant differences, correlations were computed to reveal dependencies between two items. Therefore, Spearman's Rank Order Correlation tests were carried out. Because of the large number of correlations, only the most relevant pairs are reported. What is most striking are the correlations concerning robot experience. In both conditions, people with higher **robot experience** wanted to **interact more with the real robot**, $r_s(34) = .458, p = .006$(social), $r_s(34) = .346, p = .045$(non-social). In the social condition, additional correlations were found. Participants with higher **robot experience** also rated the movements of the robot as less **disturbing**, $r_s(34) = -.343, p = .047$, thought more that the robot **can become a friend**, $r_s(34) = .505, p = .002$, **trusted the information of the robot** more, $r_s(34) = .441, p = .009$ and thought more that **the robot can act on its own**, $r_s(34) = .397, p = .020$. In the non-social condition, two additional correlations appeared. Participants with higher **robot experience** rated the movements of the robot as more **unpleasant**, $r_s(34) = .454, p = .007$ and could better **estimate the behavior** of the robot, $r_s(34) = .358, p = .038$. Participants that **tended to criticize others** rated the robot as more **unsocial**, $r_s(34) = -.406, p = .017$ and less **reliable**, $r_s(34) = -.417, p = .014$ in the non-social condition. In the social condition this correlation disappeared.

8. STUDIES

When looking into the qualitative data, the following main statements were extracted. In the non-social condition, one participant stated that the speech was too fast. In the social condition all four comments stated that the robot performed natural and human-like movements.

8.1.4 Discussion

The aim of this study was to test whether the combination of different DBCs has similar effects on a human participant as single DBCs. Therefore, a small animation was created that used different DBCs that all had similar effects when rated on their own. As stated in section 3.1 combining different signs does not necessarily result in a combination of its effects because of interferences. Here, we investigated whether negative interference appeared with the chosen DBCs. In addition, this study was used to test if the first and second hypothesis introduced in section 3.3 could be supported. Although the functional behavior of the robot was the same in both conditions, the small differences that were generated by using DBCs were capable of influencing how the user perceived the robot. Because of this, the foundation of these two conditions can be used for the next study to test if these influences can also affect arousal and the task performance of the human (Hypothesis 3).

The second Hypothesis says that DBCs are able to build different character traits for the robot. Therefore the functional behavior is compared with a social behavior that uses all DBCs to test if this shifts the perceived robot trait in similar directions than the single ones. The items that belonged to the first group of the questionnaire were used whose evaluation showed that the second Hypothesis can be supported. In this group, 9 of 14 items show significant results, and in all significant cases, the robot trait was rated as very positive and friendly as intended by the chosen DBCs. For example the robot was rated as more attractive, emotional, social, natural and likable compared with the non-social condition. An interesting effect occurred concerning the perceived intelligence of the robot. There were no significant differences for the item dense/intelligent and unreliable/reliable. This can be interpreted such that social behavior has no influences on the perceived functional features of the robot. Indeed, the item incompetent/competent showed significant differences but the reason for this might be that competence itself has a social component. This would mean that there is a decoupling between the perceived functional and social features where changes in the functional behavior only influence the perception of functional features, and changes in the background behavior only influence social perception. In general the robot was rated as very attentive, relaxed and calm in both conditions. Even though there were no significant differences, these are character traits that support the desired robot personality for further research. To sum up, it can be said that the combination of different DBCs functioned as intended and are able to build character traits as desired. Future research should evaluate how the perception of the robot changes if DBCs are used that do not fit together.

Analogous to the previous studies, this study was used to test if using DBCs increased the familiarity and liveliness in the robot (Hypothesis 1). It could be shown that combined DBCs can also support this Hypothesis. Here, the items of the second and the third group were used in combination with the qualitative data. First, the movements in the social condition were rated as very positive over all items. For example, they were rated as elegant and pleasant. This shows that the participants liked them , and this is a sign of familiarity. In addition, the ratings of the third group confirmed the high familiarity of the robot. The results were statistically significant when showing that participants were more likely to want to make friend with the robot, feel closer to the robot, can estimate the behavior or trust the robot more. The positive ratings for the items "How vivid did the robot appear to you? " and "How human-like appeared the robot to you? " appear to support the second part of the first hypothesis which argues that DBCs increase perceived liveliness in robots. One interesting result that does not directly belongs to the first hypothesis is that in the social condition the participants believe more strongly that the robot was capable of autonomy which shows that people anthropomorphise the robot more when it behaves like a person. In both conditions, the robot acts on its own, but in the social condition the expectations are higher that it does so. When evaluating the qualitative data, the same trends can be seen. All comments that were given in the social condition confirm that the behavior of the robot is perceived as very natural and human-like. All these data show that a robot can be greatly improved on a social level when using DBCs instead of only functional signs. All of these adaptations of human background behavior make the human participant feel closer to the robot, which most likely results from the decreased expectation mismatches.

8. STUDIES

Beside supporting the hypotheses, correlations between the items and the personal data of the participants were evaluated. The most prominent correlations concerned the robot experiences that the participants had. They show that robot experiences influence the perception human participants have towards a robot. Knowledge about the mechanics of a robotic system and its capabilities in general can give clues as to the capabilities of a specific system, and therefore can be influential, especially when using deceptive social cues. In both conditions, participants would like to interact more with the real robot when they have more experience with robots. This could be caused by a more general interest in robotic systems once some experience with robots has been gained. An interesting effect is that participants with more robot experience seem to anthropomorphise the social robots more (treat it like a friend, trust the robot more) while anthropomorphising the non-social robot less (movements are more unpleasant). This could be explained by expectations the participants gained in past interactions with other robots, which may somehow overlay the human expectations onto the system. They know that robots have richer capabilities than just functional behavior, and are able to distinguish between real robots and fictional robots presented in movies and literature. Another correlation that supports this further is the fact that participants with higher robot experience are able to better estimate the movements of the robot because they could be familiar with the movement patterns used for functional robots. Other prominent correlations are those of people that tend to criticize others. In the non-social condition, they rated the robot as less reliable and less social compared to participants that did not tend to criticize others. This could mean that the non-social robot is already anthropomorphised enough such that people rate it as human, and therefore become the target of criticism. The more interesting effect is that in the social condition, this correlation vanishes meaning that even when people criticize a lot, there is no reason to criticize more in the the social condition. This again shows that the social condition is familiar to the users.

To summarize, the behaviors generated with DBCs can reasonable enrich the functional behavior of a robot. They can generate a robotic personality and can make the robot appear more lively and familiar by only providing background cues as was hypothesized in this thesis. Based on this, it is possible to test the third hypothesis using the set of DBCs tested in this online study.

8.2 Interaction Study

In this section the interaction study is described that was used to evaluate the third Hypothesis. The participants had to interact with the real iCub in a laboratory. During this study, the robot functioned as the experimenter itself. During the interaction, it gave all instructions concerning the tasks the participants had to perform, and led the whole interaction. The robotic character traits that were used for this study were generated by the DBCs that where evaluated during the previous online studies. Thereby those parameter were used which create a friendly robot to facilitate a pleasant cooperation between the robot and the participant. To evaluate this study, several measurements were used. Beside questionnaires and video data, biophysiological measurements were used to record the arousal level of the participants. This was done to test the third Hypothesis which says that "minimizing the amount of expectation mismatches by using DBCs decreases the arousal level of a human user within a cooperative task". Using biophysiological measurements is an objective way to gather data within an interaction because the participant is not able to influence this data willingly. Therefore, these results allow direct access to the internal mental states of the participant. Beside this data, task performance was evaluated by analyzing the duration the participants needed to solve the given tasks. Two conditions were compared in this study. In the first condition, the robot only provided functional behavior to explain the tasks. In the second condition this behavior is enriched by using DBCs.

8.2.1 Setup

This study was designed as a between-subject study where participants took part in only one of the two conditions. During the interaction, the participants sat in front of the robot at a table on which they were asked to perform some simple tasks such as stacking bricks and remembering words. After connecting all sensors that where needed to gather the biophysiological measures, the whole study, including welcoming, giving instructions and heading over to the questionnaire at the end, was moderated by the robot. The participants simply had to respond to the behavior of the robot in this face-to-face interaction. Compared to the proof of concept study

8. STUDIES

(see chapter 5), some major changes were done in the study design to take into account the criticism made of the first study. First, the robot was no longer passive. It actively led the conversation, and in addition was able to speak and gesture. Second the DBCs that were used were improved as described in chapter 6. Lastly the design of the tasks was chosen such that the participants had to focus more on the robot and not on the human experimenter.

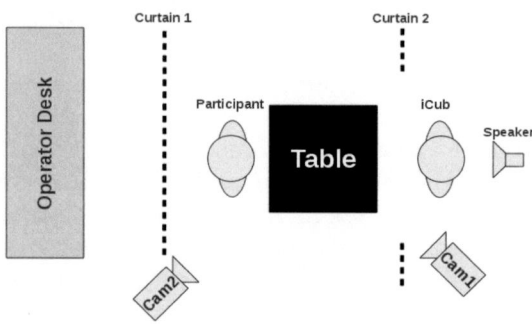

Figure 8.1: Schematic diagram of the setup used in the interaction study

During the interaction with the robot, the participants were sitting at a table. The distance of the chair and the table was always nearly the same. The left arm of the participants were used to attach both sensors. As such, movement was not allowed in that arm to avoid corrupting the data. The ESR sensors were attached to the index and middle fingers and the ECG sensor was attached to the ring finger. To make this as comfortable as possible, and to not inhibit blood flow, a soft layer was placed on the table to rest the arm on. The whole table was covered with a black blanket with a hole in the middle (the center area) where the bricks could be stacked during the tasks. For the tasks, the participants were only allowed to use the right hand. On the right side of the table, the bricks were placed that were needed to build the towers. On the opposite side of the table, the iCub was placed so that it could be seen from the knees upwards (compare Figure 8.3(right)). The distance of the robot and the table was chosen so that the robot was not able to hit it. To cover the robot when necessary, a curtain was put between the robot and the table that could be closed and opened. Another curtain was placed behind the participants to cover the operator desk. Thereby, a relatively thin material was chosen. Because of this, the operator was able to observe the illuminated interaction area while the participant was not able to see through the curtain in the other direction into the darker operator area. To illuminate the area of interaction, normal ceiling lights were used. In addition, two softboxes illuminated the face of the participants to enhance face recognition. The use of softboxes and the high position of them prevent participants from being blinded. To generate the speech of the iCub, a little speaker was placed behind the robot where the participants could not see it. To record the scene, two cameras were used which were placed in such a way that they did not disturb the participants very much. One camera was placed behind them where it was out of sight, and the other behind the iCub beneath the softbox where it was occluded as much as possible from the participants viewpoint, without interfering with the recorded image. Aside from the images of the camera, all information that were inserted into the `SelfAwareness` memory was saved to reconstruct the movements of the robot later on.

Figure 8.2: Possible solutions for the three building tasks with increasing difficulty from left to right.

In this study, the participants had to solve 6 tasks in total. Each task was explained by the robots, who waited after the explanation until the task was finished, and then began explaining the next task. The tasks were divided into two blocks. The first block contained motoric tasks, and the second block cognitive tasks. These types were chosen to challenge the participants in two different ways because it is possible that the social behavior of the robot causes different effects in both types of tasks. Both blocks were presented in random order to prevent order effects. The tasks

8. STUDIES

within the groups were designed to be similar but with increasing difficulty. In the motoric block, the tasks were chosen to build towers out of little wooden bricks. In the first task the participants were told to take 10 bricks and build a tower that was as stable as possible. The second task was to build a tower with 10 bricks that was as small as possible with no brick lying beside the other. The last task was to take four bricks and to build the highest possible tower. A small pretest showed that four bricks is challenging but possible for nearly all tester. The challenging task should not be too challenging so as to not generate such a high level of stress that overshadowed the effects caused by the robot. The sample solutions for all three tasks is shown in Figure 8.2. The participants did not have to reach the solutions shown in the examples. As soon as they signaled that they had finished the task, it was marked as solved. To not generate additional stress if a tower fell over, the table was covered with a soft layer to dampen the sound produced by falling bricks. The tasks of the second block involved remembering objects such as house, key or telephone. In the first task there were 3 items, in the second 5, and 7 items in the most challenging task.

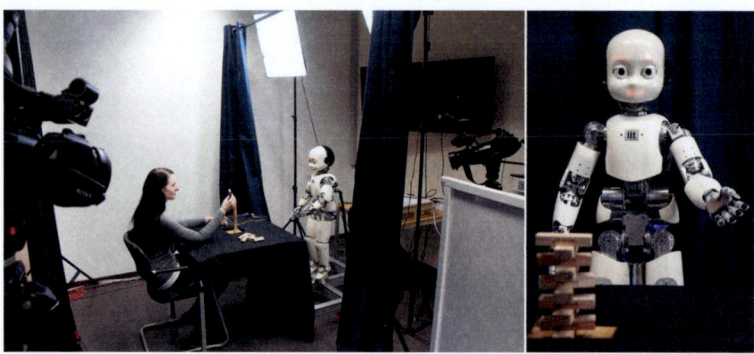

Figure 8.3: Picture of the setup used in the interaction study. Left: Picture of the whole setup. Right: iCub from the participants view during the study

Participants had to pass multiple stages during this study. After welcoming them, the human experimenter showed them the robot so that every participant had an idea of what they would be interacting with. After that, each participant had to fill out a pre-experimental questionnaire. In parallel, the experimenter closed the curtain that covered the robot from the view of the participant so that the robot's appearance did not disturb them during the pretests. After finishing the questionnaires, the sensors were attached to the participant's left hand followed by two pretests. As described in the next section, these were necessary to evaluate the biophysiological measures. After these tests, the curtain was opened, making the iCub once again visible. Now, the behavior depending on the condition was activated and all recordings were started. After that, a beeping sound was played for the cameras, and a marker signal was inserted into the data streams. This beep and the marker were added at exactly the same time and were used to synchronize all data streams later.

After all these preparations, the robot interaction began. The participants were told to pay attention to the robot, and to behave as they thought appropriate. It should mentioned that no instructions were given by the experimenter. All this was done by the robot. The interaction with the robot was a longer conversation consisting of several phases. To display the conversation, the robot used speech, and background behavior as well as functional gestures (compare Figure 8.4). Since no feedback on user input was required, to help keep the interactions more comparable, the order and timing of speech and gestures was fixed. There were some points in the interaction where the robot waited until the task was solve. To recognize the end of the task this was triggered by the operator. The first phase was the welcoming phase. Here the robot introduced itself and thanked the participants for their participation. In the second phase, the robot explained that some tasks had to be solved and that the participants should take care to not move the arm with the sensors. During all introductions the robot always used adequate gestures, i.e. pointing onto the sensors or displaying a welcome gesture, as it will be described below. In the next phase, one of the two task blocks was carried out. It began with general instructions and followed with instructions for the individual tasks. While the robot was waiting for the participant, it either gazed at the center area during the building tasks, or used the gaze DBCs to focus in the participant's face during the cognitive tasks. After each task, the robot showed a neutral reaction to keep the behavior of the robot as similar as possible for all participants. Common reactions were: "OK." or "Now we finished the building task.". In the last phase, the robot thanked the participants again for their participation and handed over to one of the human experimenters. After this interaction, all participants were asked to fill out a second questionnaire containing post-experimental tests and personal data. During the entire study including the first presentation of the robot when DBCs

8. STUDIES

could not be seen, the iCub was put in the posture corresponding to the condition the participant was in to help keep the perception of the robot consistent during the complete study. For the same reasons, the power supply of the robot was kept running the entire time, including after the interaction was completed, while the participants fill out the questionnaire. The whole sequence of the robot interactions is presented in Appendix B.6 as a translation from the original German instructions.

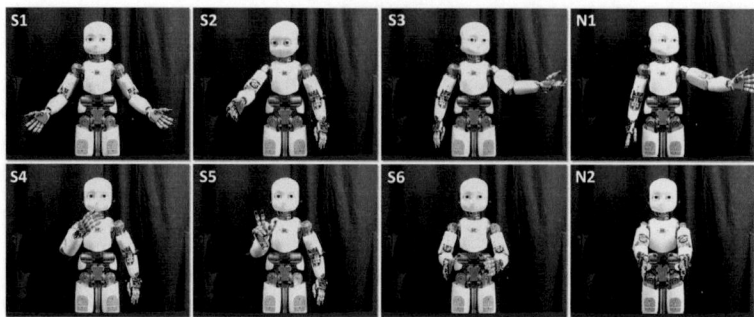

Figure 8.4: Different gestures produced during the interaction Study. S = social condition; N = non-social condition

During this study, two conditions were used. First, the non-social only used functional behavior. Second, a social condition displayed background behavior using DBCs as well as functional behavior. The behavior displayed by the robot was generated autonomously. The conversation and the gestures used were preprogrammed to be consistent across participants. Gazing behavior was generated by using the face recognition and saliency module. The only input given by a person during the study was the trigger when the participant finished one task so that the robot could move on with the conversation. The only difference between the conditions were the gestures produced by the robot, described as follows. Appendix B.6 lists the gestures used and their timings. Each row represents one PiRA-XML sent by a tool of the `Functional Layer`. In addition Figure 8.4 shows some sample gestures. To design these, four people were recorded engaged in the same conversation the robot would reproduce in the study. This was done to get an impression of how people behave in this situation. The behavior of the robot was based on these videos.

In the non-social condition, the only behavior that was displayed was the functional behavior required to complete the task. The main sign used was speech. To prevent displaying background behavior, no lip movements were shown. In addition, the gestures were reduced to their functional core. For example, to point in a given direction, only the shoulder joint was used to move the arm (Figure 8.4(N1)). The rest of the arm was kept immobile because it was not needed to display the functional sign. Another example is the gesture used to clarify the term for "stability" where only the shoulders were used to move the hands in front of the robot (Figure 8.4(N2)). The gazing behavior was generated by only using the neck joints. Here, no eye-neck coordination was used because this is a background cue. Gaze direction was all that remained after removing the background signs.

In the social condition, the functional signs were exactly the same as in the non-social condition. The only difference was that these were executed in a more social way. For example, the pointing gesture in Figure 8.4(N1) looks different than in the social condition (Figure 8.4(S3)). To move the arm into the posture in the picture a natural movement (see section 6.2) was chosen that used all the joints of the arm to make it more dynamic. The functional information for both gestures was the same but on a social level the gesture transmitted much more. In this condition, all movements were modified to be more natural compared with the non-social condition. Furthermore, some additional gestures were included strictly as idly movements such as lifting the arm during the greeting (Figure 8.4(S1)). Figures 8.4(S1-S6) show more examples of other gestures that were executed in the social condition, whereby gesture S6 is the same as N2 but in the other social condition. Aside from the natural gestures, the lips were used to support speech, and eye-neck coordination and gaze DBC were used to make the gazing more natural. In addition, the breathing DBC was used for idle movements between the animations. All parameterized DBCs were configured such that they facilitated a friendly robot personality as was evaluated in the whole body study (see section 8.1). All of these additional cues generated a naturally moving, more socially behaving robot could influence human perception without changing the functional signs. This is exactly the effect that will be evaluated the third Hypothesis.

8.2.2 Measurements

To evaluate this study, several measures were used as described in this section. First pre- and post-experimental questionnaires were used to evalu-

8. STUDIES

ate changes in self-assessment during the study. Furthermore, participants were asked to rate the robot using existing item sets such as the Godspeed Questionnaire. In addition, objective data were gathered using biophysiological measures and by evaluating video data.

8.2.2.1 Questionnaire

The questionnaires used are divided into pre- and a post-experimental questionnaires. This was done to document possible differences before and after interacting with the robot. The questionnaire adopted here the Positive and Negative Affect Schedule (PANAS), a self-assessment questionnaire designed by Watson et al. [159] used to evaluate the participants mood. The extended version of this test (PANAS-X [158]) consists of 60 items, although only 28 were used. Each item can be sorted into several categroies, i.e. Positive Affect, Negative Affect, Fear and Joviality and was evaluated on a 5 point scale ranging from "very slightly or not at all" to "extremely". The items utilized are the following:

- **Negative Affect:** afraid, scared, nervous, jittery, irritable, hostile, guilty,
  ashamed, upset, distressed.

- **Positive Affect:** active, alert, attentive, determined, enthusiastic, excited, inspired, interested, proud, strong.

- **Fear:** afraid, scared, frightened, nervous, jittery, shaky.

- **Joviality:** happy, joyful, delighted, cheerful, excited, enthusiastic, lively, energetic.

For this evaluation, we administered the German version of the test as described in [131]. The test was completed twice to see if participants rate their own mood differently after the interaction with the robot. If they liked the interaction, it was expected that their mood would change accordingly. A similar procedure was carried out for example by [89]. In addition to the PANAS test, some questions concerning the biophysiological measures were completed in the pre-experimental questionnaire, as described in the following section.

The post-experimental questionnaire consisted of five smaller questionnaires. The first was the second PANAS test. To compute the score for each item, we used the difference between the first and the second PANAS. The second questionnaire was the Godspeed Questionnaire invented by Bartneck et al. [6]. It was created to provide a standardized means of rating robots for HRI. It consists of 24 items divided into five groups. Each item is rated on a 5 point scale ranging from one extreme to the other, i.e. machinelike/humanlike. For this study, only the following four groups were used: Anthropomorphism, Animacy, Likeability, and Perceived Intelligence. For each group, a score was computed to generate a general value. This was done by summing up the scores of every item in the corresponding group. The whole questionnaire can be found in the Appendix Figure A.14. The questionnaire was used to rate the general perception of the robot depending on the condition. The third questionnaire was another PANAS questionnaire but this time the participants had to rate what they thought of the mood of the robot, i.e. what emotions the participants ascribed to the robot. Here also scores were computed to generate values for each category such as negative affect, positive affect and so on. To this end, each score of the contained items was summed up. An additional score was the difference between the positive and negative affect as computed to evaluate whether there were differences between the two conditions. The fourth questionnaire was the Big5 scale modified so that the questions asked were about the robot and not the participant. For example, the item "I am diffident, reserved." was changed to " iCub is diffident, reserved." A more detailed description can be found in chapter 5. This questionnaire was added to evaluate how the participants rated the robot's personality.
The first four questionnaires focused mainly on social aspects. In contrast the last questionnaire contained items that rated the general interaction on a more functional level. For example, participants rated which task was most difficult and which was most fun, if the interaction was pleasant or if they felt stressed. Mainly these items were taken from a questionnaire developed by Parise et al. [119] to rate games. We used it to have participants rate the tasks they had to complete. In addition, some items were taken from the online studies described above, i.e. "Do you trust the information given by the robot? ". All items were rated on a five point

8. STUDIES

scale. The complete questionnaire can be found in Appendix Table A.1.

After completing the post-experimental questionnaire additional personal data was collected. This questionnaire was the same as for the whole body study described in section 8.1 and requests age, gender, native language, job, highest degree earned, computer experience, robot experience and the Big5 of the participant. Lastly participants had the option to comment on the measure and the study in general.

8.2.2.2 Biophysiological Measurement

Within this study the *biophysiological measures* (BioMs) ESR and ECG were used to measure the arousal of the participants. To gather data that could be evaluated, several prior steps were necessary, and several specifics had to be considered. The basic theoretical background for BioMs was already discussed in section 2.4. Therefore only study relevant differences are described here. The main BioM used was *Electrical Skin Resistance* (ESR). The *Electrocardiography* (ECG) was used to validate the first measure. In total four steps were necessary to collect all relevant data from the participants. The first three steps were all pretests essential to interpret the main data. The first was a questionnaire that asked about the consumption of goods that could influence the measurements. This was followed by a baseline recording when the participants are relaxed, and a reference measure when they were at medium arousal.

As discussed by Bethel [9] a change in arousal is reduced when the same stimuli is presented repeatedly. In this study no stimuli were presented twice; all robot instructions differed and no animation was used more than once, and therefore there is no habituation on one stimuli

Testing of the hypothesis relied on possible differences in arousal, making this the most important measure. The used method has been used to evaluate stress induced by driving in different areas [67]. The single reactions for certain stimuli have not been evaluated yet.
The data were normalized to address the individual differences between participants. A common approach is to substract the baseline measure of the participant to compare only the increase in arousal during the stimuli presentation [143][73][67]. The problem remains, however, that even after normalizing the data some people react more or less strongly to the same

stimuli. Healey et al. [67] used the variance of the baseline to normalize all data. This means that when the participant showed stronger variations during the baseline recordings, the main data were normalized with this factor to compensate for these individual differences. To achieve a similar effect in this study, a reference measure was taken of the arousal during a standard task all participants had to do. This allowed us to get an insights into the tendency of the participants to react more or less strongly to a stimulus, and to normalize the interaction data. The formula to compute the final score used to test for significant differences between the conditions was $Score = (mean\_reference - mean\_baseline)/(mean\_interaction - mean\_baseline)$. This computed the ratio between the reference measurement and the interaction. The Score represented the relative increase in arousal as compared to the reference value. If this value was positive, the arousal of the interaction was higher than the arousal of the referent. The following example will clarify this correlation. In Figure 8.5 a typical arousal plot for the complete study is shown. The three segments on the x-axis are the three measures used to generate the score. Out of each plot, for a single segment a mean value can be computed (red lines), resulting in three values, one for each segment. The difference between the baseline and the reference values can be used for normalization because they show how differently people react to the same stimuli. Out of these values, the score can be computed using the given formula.

The pre-experimental questionnaire contained items to help interpret unexpected ESR or ECG data. BioMs can be influenced by caffeine, nicotine, medicine and drugs as well as if the participants had eaten something just before the study. All participants were instructed to not consume alcohol before the study. Furthermore, the participants were instructed to drink coffee or smoke as usual in the hours before the study to not evoke deficiency signs. In the pre-experimental questionnaire, we asked how much time had passed since they last consumed nicotine, caffeine, alcohol and food. In addition, we asked participants if they were taking any medication.
To measure the baseline, the participants were instructed to relax for 5 minutes. During this time the participants had to watch some pictures so that they were not distracted by any arousing thoughts. The picture book that was provided consists of several calming images. They were mainly landscapes of forests or beaches, flowers or animals. While looking at the book the BioMs of the participants were recorded as the baseline.

8. STUDIES

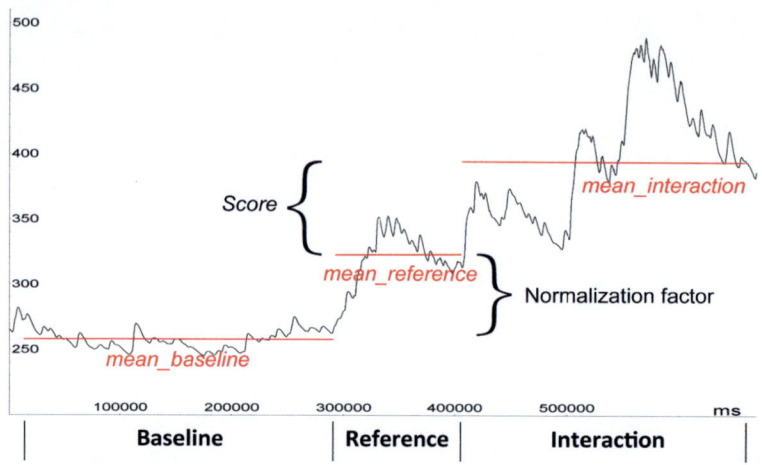

Figure 8.5: Example computation of the score needed to compare the ESR data

For the reference normalization measure, we needed to measure arousal during a task with a relatively strong ESR response. We chose holding the breath for precisely the reason. In addition, the task can be used to ensure that the participant had normal ESR, and it is a task anyone can perform. Each participant was instructed to hold their breath for several seconds, and to repeated the procedure three times.

After conducting all these pretests, the main study took place. As described above, the robot was uncovered again and the interaction began while recording BioMs during the whole interaction. To better interpret the reactions of the participant on each stimuli it was also recorded in the ESR data when each stimuli was presented. In this evaluation, the mean arousal for the entire interaction was used and normalized with the previously collected data.

8.2.2.3 Video analysis

The other objective measures used to evaluate the study were the video data. Two video streams were recorded to examine the performance.

As described in section 8.2.1, two cameras recorded the scene from two different angles. One from behind so that the back of the participant and the front of the robot was visible. The other from the front so that the back of the iCub and the participant including the table were visible. For the performance evaluation, only the front view was used because it showed the participant's behavior.

To measure task performance both, task blocks were evaluated. For the building task, the time needed to complete each task was taken to test for significant differences between conditions. Task duration was operationalized as the time elapsed from the end of the robot's instruction until the participant released the last brick. For the memory task we counted the number of correct answers the participants remembered out of the given items. In both cases the video data was analyzed using manually annotation.

8.2.3 Results

The results of this study are described as follows. 47 participants (27 female, 20 male) participated in this study. Their ages ranged from 20 to 68 years ($M = 30.47$, $SD = 11.21$). The participants rated their own computer experience on a scale from 1(none) to 5(very much) as $M = 3.45$ ($SD = 1.04$) and their robot experience as $M = 1.68$ ($SD = 0.7$). 23 participants took part in the non-social and 24 in the social condition.

8.2.3.1 Questionnaire

Given the large amount of data, only the most striking results are reported. All relevant data can be found in the Appendix B.7. First the data gathered by the PANAS for relative self-assessment were evaluated to test if the participants experienced a change in affect after taking place in the study. Because of comparing the differences between both questionnaires it has to be assumed that the rated variables are interval variables. Using t-tests, we found no differences for the item **afraid** between the non-social ($M = 0.09$, $SD = 0.596$) and the social ($M = -0.13$, $SD = 0.947$) condition, $t(45) = 0.914, p = .366, d = 0.278$. Furthermore there was a slight trend that participants felt more **lively** after the interaction in the social condition ($M = 0.04$, $SD = 0.859$) than in the non-social condition ($M = -0.52$, $SD = 1.275$), $t(45) = -1.784, p = .081, d = 0.515$. Participants in the social condition ($M = 0.25$, $SD = 0.676$) felt significantly more **proud** than in the non-social condition ($M = -0.26$, $SD = 0.915$), $t(45) = -2.183, p = .034, d = 0.634$. There was no significant difference for the item **cheerful** between the non-social ($M = $ -

0.13, $SD = 0.869$) and the social ($M = 0.08$, $SD = 0.584$) condition, $t(45) = -0.994, p = .326, d = 0.284$. In addition there was a trend that participants of the social condition ($M = -0.08$, $SD = 0.974$) were more **determined** after the interaction than in the non-social condition ($M = -0.61$, $SD = 0.941$), $t(45) = -1.879, p = .067, d = 0.553$.

For the Godspeed questionnaire we report the following results. To evaluate the single items, Mann-Whitney Tests were used. There was no significant difference for the item **artificial/lifelike** between the social ($Mdn = 3$) and the non-social ($Mdn = 3$) condition, $U = 195 p = .071, r = 0.263$, whereas the trend points in the direction that the social condition is rated as more lifelike. In addition the participants rated the robot as significantly more human-like in the social ($Mdn = 3$) than in the non-social ($Mdn = 2$) condition for the item **machinelike/humanlike**, $U = 181.5 p = .036, r = 0.306$. For the item **fake/natural** there is a strong trend that the robot is seen as more natural in the social ($Mdn = 3$) than in the non-social ($Mdn = 3$) condition, $U = 189.5 p = .054, r = 0.282$. Furthermore there is another trend for the item **stagnant/lively** which shows that the social condition ($Mdn = 4$) is rated as more lively than the non-social condition ($Mdn = 4$), $U = 191 p = .056, r = 0.28$. Lastly, the robot is rated as more pleasant in the social ($Mdn = 4.5$) than in the non-social ($Mdn = 4$) condition for the item **unpleasant/pleasant**, $U = 188.5 p = .04, r = 0.3$. For the scores of the Godspeed questionnaire the following results emerged. T-Tests were used to evaluate the data. There was a significant difference between the social ($M = 15.38$, $SD = 4.372$) and the non-social ($M = 13.09$, $SD = 2.968$) condition. Here the social condition was **anthropomorphised** more, $t(45) = -2.09, p = .042, d = 0.613$. There was no significant difference for **animacy** between the non-social ($M = 18.13$, $SD = 4.445$) and the social condition ($M = 20.13$, $SD = 4.1$), $t(45) = -1.6, p = .117, d = 0.468$. In addition there was no difference between the social ($M = 21.42$, $SD = 2.165$) and the non-social ($M = 20.48$, $SD = 3.527$) condition for **likability**, $t(45) = -1.104, p = .275, d = 0.321$. For **perceived intelligence** there was no significant difference between the non-social ($M = 18.83$, $SD = 2.774$) and the social ($M = 18.96$, $SD = 3.983$) condition, $t(45) = -0.132, p = .896, d = 0.038$.

The combined scores for the Big5 questionnaire showed the following results. The results of the single items are presented in the Appendix Table B.11. Because of the fact that scores are computed it has to be assumed that the variables are interval variables and therefore T-Tests are carried out. There was no significant difference for **extraversion** between the non-social ($M = 5$, $SD = 1.931$) and the social ($M = 5.92$, $SD = 1.666$) condition, $t(45) = -1.745, p = .088, d = 0.51$. In addition there was no difference between the non-social ($M = 7.17$, $SD = 1.302$) and the social ($M = 7.21$, $SD = 1.285$) condition for **agreeability**, $t(45) = -0.091, p = .928, d = 0.03$. For the item **Neuroticism** there is no difference between non-social ($M = 8.52$, $SD = 1.377$) and social ($M = 8.88$, $SD = 1.035$), $t(45) = -0.997, p = .324, d = 0.296$. A significant difference occurred between the non-social ($M = 5.83$, $SD = 1.23$) and social ($M = 4.83$, $SD = 1.903$) condition for **openness**, $t(45) = 2.113, p = .04, d = 0.624$. At last the item **Conscientiousness** revealed no significant differences between non-social ($M = 7.87, SD = 1.792$) and social ($M = 8.38$, $SD = 1.313$), $t(45) = -1.107, p = .274, d = 0.325$.

To evaluate the PANAS the participants filled out to rate the affects of the robot, Mann-Whitney Tests were used. The most interesting results were the following. There is a significant difference for the perceived **activeness** of the robot where the social ($Mdn = 4$) condition was rated as more active than the non-social ($Mdn = 3$) condition, $U = 185 p = .04, r = 0.229$. Furthermore the social ($Mdn = 4$) condition was rated as more **alert** than the non-social ($Mdn = 3$) condition, $U = 166.5 p = .016, r = 0.352$. For the combined scores the following results were produced using T-Tests. There was no difference for the **negative affect** rating between the non-social ($M = 11.7$, $SD = 3.066$) and the social ($M = 11.42$, $SD = 2.062$) condition, $t(45) = 0.367, p = .715, d = 0.107$. The social condition ($M = 30.92$, $SD = 6.554$) was rated as having more **positive affects** compared to the non-social condition ($M = 26.65$, $SD = 7.901$), $t(45) = -2.018, p = .05, d = 0.588$. Furthermore the **difference between Positive and Negative Affect** was more distinct in the social ($M = 19.5$, $SD = 6.108$) than in the non-social ($M = 14.96$, $SD = 8.626$) condition, $t(45) = -2.091, p = .042, d = 0.607$.. There were no differences between the non-social ($M = 7.39$, $SD = 2.251$) and the social ($M = 7.08$, $SD = 1.666$) condition for **fear**, $t(45) = 0.535, p = .595, d = 0.157$. Lastly, there was no difference for **joviality** between the non-social ($M = 17.52$, $SD = 7.179$) and the social ($M = 20.33$, $SD = 6.638$) condition, $t(45) = -1.395, p = .17, d = 0.406$..

The last questionnaire, containing the functional questions, consists of thirteen questions that were rated on a five point scale. The evaluation was done by performing several Mann-Whitney Tests. For the question **"How easy/difficult were the tasks?"** the social condition ($Mdn = 2$) and the non-social condition ($Mdn = 2$) did not differ, $U = 271.5, p = .919, r =$

8. STUDIES

0.015. There is no difference for the question **"To what extend did you feel stressed?"** between the non-social ($Mdn = 2$) and the social ($Mdn = 2$) condition, $U = 270.5, p = .899, r = 0.019$. For the question **"Did you feel uncomfortable?"** no differences appeared between the non-social condition ($Mdn = 2$) and the social condition ($Mdn = 2$), $U = 257.5, p = .675, r = 0.061$. There occurred a significant difference for **"Was it fun to interact with the robot?"** which was rated as more fun in the social ($Mdn = 4.5$) than in the non-social ($Mdn = 4$) condition, $U = 192, p = .048, r = 0.288$. In the social ($Mdn = 5$) as in the non-social ($Mdn = 5$) the participants gave a high rating for the question **"How willing would you be to repeat the interaction?"**, although there was no difference between the conditions, $U = 250.5, p = .497, r = 0.099$. There was no difference for the question **"How interested were you?"** between the non-social ($Mdn = 4$) and the social ($Mdn = 5$) condition, $U = 230, p = .272, r = 0.16$. Furthermore there was no difference for the question **"How attentive were you?"** in non-social ($Mdn = 4$) as well as social ($Mdn = 4.5$), $U = 221, p = .207, r = 0.184$. For the item **"How happy were you?"** participants in the social condition ($Mdn = 4$) were more happy than those in the non-social condition ($Mdn = 3$), $U = 169, p = .011, r = 0.37$. There was no difference between social ($Mdn = 4$) and non-social ($Mdn = 3$) for the question **"How satisfied were you?"**, $U = 214, p = .157, r = 0.207$. For the question **"Do you trust the information given by the robot?"** there was also no difference between social ($Mdn = 4$) and the non-social ($Mdn = 3$) condition, $U = 205, p = .116, r = 0.23$. Furthermore there was no difference for the item **"Would you like to interact with the robot more often?"** between the non-social ($Mdn = 4$) and the social ($Mdn = 5$) condition, $U = 211, p = .145, r = 0.213$. For the question **"Are you able to appraise the behavior of the robot?"** there was no difference between the non-social ($Mdn = 2$) and the social ($Mdn = 3.5$) condition, $U = 225.5, p = .256, r = 0.166$. For the last question (**"Do you think the robot can act on its own?"**) there was no difference between the non-social ($Mdn = 2$) and the social ($Mdn = 3$) condition, $U = 222.5, p = .241, r = 0.171$.

Beside the differences, correlations between the items and the personal data were evaluated. We used Spearman's Rank Order Correlation. Here only the most relevant correlations are presented. It turned out that mainly three personal data items revealed the most important correlation. These are the age of the participants, computer experience and robot experience. Participants with a higher **age** rated the robot in the non-social condition as more **competent** ($r_s(23) = .535, p = .009$), **sensible** ($r_s(23) = .449, p = .032$), more **conscious** ($r_s(23) = .595, p = .003$), **responsible** ($r_s(23) = .510, p = .013$), **intelligent** ($r_s(23) = .523, p = .010$), **organic** ($r_s(23) = .444, p = .034$), having more **perceived intelligence** ($r_s(23) = .657, p = .001$) and they **trusted information from the robot** more ($r_s(23) = .707, p = < .001$). In the social condition participants with a higher age rated themselves after the interaction as being less **stressed** ($r_s(24) = -.49, p = .015$), less **uncomfortable** ($r_s(24) = -.628, p = .001$), and more **satisfied** ($r_s(24) = .417, p = .043$).

Those participants with greater **computer experience** rated the robot in the non-social condition as less **humanlike** ($r_s(23) = -.562, p = .005$), more **inert** ($r_s(23) = -.506, p = .014$) and having less **animacy** ($r_s(23) = -.417, p = .048$). In the social condition these participants rated the robot as more **inert** ($r_s(24) = -.496, p = .014$), more **foolish** ($r_s(24) = -.418, p = .042$), **ignorant** ($r_s(24) = -.417, p = .043$), having less **animacy** ($r_s(24) = -.416, p = .043$), being less **alert** ($r_s(24) = -.549, p = .005$), and less **positive** ($r_s(24) = -.453, p = .026$).

Further correlations occurred for the item **computer experience**. With increasing experience the robot was rated in the social condition as more **humanlike** ($r_s(24) = .409, p = .047$), more **pleasant** ($r_s(24) = .433, p = .035$), having a higher **likability** ($r_s(24) = .509, p = .011$) and being **stronger** ($r_s(24) = .426, p = .038$). In addition these participants rated themselves as **happier** ($r_s(24) = .425, p = .038$) in the social condition and would **like to interact with the robot** more often in the social conditions, $r_s(24) = .574, p = .003$ and less often in the non-social condition $r_s(23) = -.423, p = .055$.

The questionnaires contained the option to give comments. First, we asked which tasks were easy or difficult. Comments on this questions confirmed the classification of the tasks as being of increasing difficulty. In Addition, the memory tasks were seen as more difficult than the building tasks. Further, most of the participants said that the speech synthesis was difficult to understand in some cases which made the memory task more difficult than expected. Comments concerning the biophysiological measurements showed that the way of measuring the data was acceptable and less disturbing. Furthermore many participants liked the study and found it very interesting. Unfortunately no concrete comments on different

8. STUDIES

aspects of the robot's behavior were given.

8.2.3.2 Biophysiological Measurements

The results of the BioMs were used mainly to test the third hypothesis which measure the arousal of the participants. For the evaluation of the ESR, seven participants were excluded. Some had no or abnormal ESR responses as described in section 2.4, others were excluded because of technical failures. 19 of the remaining participants took part in the non-social condition and 21 in the social condition.

To evaluate the **ESR** a Mann-Whitney Test was carried out as the test for normality failed for this parametric variable, (Shapiro-Wilk Test $p < 0.05$). For the **ESR** there was a significant difference between the non-social ($Mdn = 1.11$) and the social ($Mdn = 1.44$) condition. In the social condition the arousal of the participants was significantly higher than in the non-social condition, $U = 119, p = .029, r = 0.345$.

The ratio for the **ECG** could not be computed because of reasons described in the discussion. Instead, the difference between the mean heart rate of the baseline and the mean heart rate of the interaction were used. For evaluation a t-test was carried out. There were no significant difference between the non-social ($M = 5.19, SD = 4.928$) and the social ($M = 7.6, SD = 6.187$) condition, $t(41) = -1.407, p = .167, d = 0.431$.

8.2.3.3 Video analysis

For the evaluation of the performances mainly Mann-Whitney Tests were used, as the variables were not normally distributed except one. For the building tasks the time needed to build the tower in milliseconds, was compared between the two conditions. There was no difference between the non-social ($Mdn = 25770.5$) and the social ($Mdn = 21772$) condition for the building of the **most stable tower**, $U = 185, p = .082, r = 0.256$. For the building of the **smallest tower** there was no difference between the non-social ($Mdn = 34372$) and the social ($Mdn = 32441$) condition, $U = 216, p = .291, r = 0.156$. There was a significant difference between the social ($Mdn = 15241.5$) and the non-social ($Mdn = 20316.5$) condition for the **highest tower**. Here the participants were faster in the social condition, $U = 121, p = .002, r = 0.464$.

For the memory tasks, the number of correct answers were used to test for differences. To remember **three** items, there is no difference between the non-social ($Mdn = 3$) and the social ($Mdn = 3$) condition, $U = 241.5, p =$

$.311, r = 0.149$. Furthermore there was no difference for the **five** items between the non-social ($Mdn = 4$) and the social ($Mdn = 4.5$) condition, $U = 259, p = .905, r = 0.018$. Lastly, there was no difference between the non-social ($M = 4.5, SD = 1.263$) and social ($M = 4.21, SD = 1.503$) condition for the **seven** items, $t(44) = 0.709, p = .482, d = 0.209$.

8.2.4 Discussion

The main goal of this study was to test our third hypothesis (see section 3.3) which argues that DBCs can influence human behavior such that performance can be improved. More precisely, that the usage of DBCs decreases the arousal of a human user, influencing task performance in a positive or negative way. To test this hypothesis, biophysiological measures were taken in a study where participants interacted with iCub showing either strictly functional or full social behavior.

As the results of the BioMs show, the hypothesis could not be confirmed although the BioM data did reveal an unexpected result. Instead of decreasing the arousal of the participant, as was predicted, a more socially behaving robot increased arousal. To interpret this effect, we further investigated what type of arousal was generated. As proposed by Bethel et all. [9] the Valence/Arousal-Model suggests that arousal can be positive or negative. Although we found that arousal was increased, it was unclear which valence the arousal had. When formulating this hypothesis, it was assumed that non-social behavior generated negative arousal because of expectation mismatches. When interpreting the questionnaires it can be seen that the whole situation the participants were in was rated as positive. Self-assessment as well as the PANAS rating of the robot, were rated as very positive and less negative in both conditions. In addition, the functional questionnaire showed that the participants indicated little stressed, did not feel uncomfortable, and in fact, had fun during the interaction in both conditions. Based on these results, it can be concluded that the arousal that was measured during the interaction was positive. In other words, DBCs increase the positive arousal of the participant significantly compared to the non-social condition.

In terms of the theoretical background, the occurrence of this effect may be interpreted as follows. People seem to be more active in a social situation, likely because they feel involved. Referring to Matthews et al. [102] "arousal refers to the person's overall state or level of activity" (p.164). This means that the participants are more active in the social condition,

8. STUDIES

possibly because of the increased positive mood the situation evokes, or by the fact that the robot is anthropomorphized more. Seeing the robot as more human-like could create the need to interact more as argued by Maynard-Smith or Watzlawick who both see each action as a reaction to a previously presented signal and each action itself generates a certain reaction. Whatever might be the exact reason, humans become more positively aroused and active when interacting with a more social robot. Conversely, they become bored if the robot only performs functional behaviors like a tool without social competences.

Nevertheless, a part of the third hypothesis was supported. DBCs have the capacity to decrease expectation mismatches as has been shown by the significant increases in anthropomorphism in the social condition. The only effect that could not be supported is that DBCs do not decrease negative arousal but instead increase positive arousal.

These results raise the question of how they can be reconciled with the research by Kulić et al. [87] who argued that social behavior in a robot reduces anxiety and arousal. Within their study, the participants were in a negative situation, since they were made to feel danger of being hit by a robotic arm. Reducing anxiety in this situation means reducing negative arousal. It could be that in negative situations social behavior reduces negative arousal, and in positive situations it increases positive arousal. This would be an interesting topic for further research.

During the data recording, a problem occurred that should be discussed here. First, our intention was that the ECG data would be used to collaborate the ESR data. Unfortunately the ECG data were influenced during the reference task by disturbing effects. As already mentioned, the heart rate is influenced by different variables such as emotional state, physiological state and so forth. This means that it reacts differently than ESR, and is therefore not the optimal way to measure arousal but is still a good way to confirm the ESR data. In the case of this study, the problem was that heart rate decreases when someone holds their breath whereas ESR increases during the reference task. This effect makes it impossible to compare the ratio between the reference and the interaction tasks because the reference data cannot be used to normalize personal differences. Because of this, only the difference between the relaxed and interaction tasks could be used to evaluate heart rate. Nevertheless, the heart rate showed similar trends as the ESR despite the lack of normalization. This shows that the interpretation presented above is supported by the heart rate data.

As a secondary effect, we evaluated whether social behavior increases task performance. The assumption was that the arousal a person has, influences task performance. If the task is easy, increased arousal improves task performance. If the task is difficult, an increase in arousal impairs it (the Yerkes-Dodson Law, See section 2.3). Considering the fact that DBCs increase arousal, and since the robot is a stressor within the interaction, this means that in theory DBCs should improve task performance for easy tasks and impair performance for difficult ones. This would be a similar effect to social facilitation where humans/robots as social entities function as stressors, and influence performance in a similar way (see section 2.3.2). As the results of the task performance showed, this effect is not very strong. First, for the easy tasks (the building tasks) the participants were significantly faster in the social condition for the highest tower, in support of the theory. For the other two towers, there was no significant difference although the mean durations for both tasks were shorter in the social condition, and at least for the most stable tower, show a slight trend. The reason for this might be that the first task was too simple and too short to find a significant effect, if there was one. For the second task, there was too much variation in the execution of the task. Nearly every participant understood the instructions differently, and build different towers, making the durations difficult to compare. For the hard tasks, there are no visible effects. There may be two explanations. First, the difficulty of this task was not high enough to show an effect, possibly culminating instead in a ceiling effect. For example because of being on the peak of the graph of the Yerkes-Dodson Law. Second the described effect simply does not occur for DBCs. Further studies are required to clarify. In summary it can be said that for easy tasks, social behavior is able to improve task performance by increasing the arousal. Further testing is required to see if it would be advisable to deactivate the social behavior to not impair the performance for difficult tasks.

The previous online studies have shown that DBCs are capable of generating different character traits. In this study, the same effects appear again. Furthermore, the character traits that were chosen based on the online studies, are again produced by the social behavior in this study. The Big5 and the PANAS for the robot show that the robot is seen as significantly more open and positive compared to the non-social condition. This shows that the generated character traits are also perceivable in a real interaction with the robot. In addition, both questionnaires showed

8. STUDIES

that the robot in general was perceived as very positive in both conditions. The absolute ratings of the PANAS in particular, reveal that the items measuring Negative Affect and Fear were rated very low (Median 1) and items measuring Positive Affect and Joviality were rated very positively (Median 1-4). This shows that the iCub itself is adequate as a social robot whose perception can be further improved by using DBCs.

In addition the Godspeed questionnaire shows similar results that support the hypothesis that DBCs let the robot appear more familiar and lively (Hypothesis 1). Here the robot was rated as significantly more human-like and pleasant in the social condition. Furthermore, it was more anthropomorphised in the social condition. The most interesting effect in this questionnaire was that the same effects as in the whole body study occurred again. The items that show significances and trends in the Godspeed questionnaire and in the other questionnaires are mainly items that describe social aspects. For example, human-likeness, openness, anthropomorphism and pleasantness. In contrast, items that describe aspects of intelligence such as responsibility, knowledgeability and perceived intelligence were mostly unaffected. This suggests the hypothesis that DBCs, or more generally background signs mainly effect the social perception of the robot, and not its cognitive abilities. Inversely, could be that functional behavior mainly influences perceived cognitive abilities and not social ones. This is an interesting topic for further research.

Beside the rating of the robot, it was of high interest how the participants felt during the interaction. For this evaluation, the self-assessment and functional questionnaires were used. In general, after the interaction with the robot in the social condition, the participants felt happier, had more fun and were prouder, as compared to the non-social condition. In addition, most wanted to interact with the robot more often, independently of the condition. These results are analogous to the ratings for the robot. In the social condition, the participants felt more positive and were mostly influenced in social aspects such as fun and pride. The only question that concerns the functional aspect was if the information provided by the robot were trusted, which shows no differences. This further facilitates the hypothesis that social behavior mainly influences social perception. Parallel to the ratings for the robot, this study showed that background cues were able to make participants happier and to influence them positively (e.g.: positive arousal) which is an additional reason why this behavior is important when designing social robots.

The evaluation of the personal data revealed three main correlations. These were the age, computer and robot experience. For younger participants, the robot was seen as having less cognitive abilities in the non-social condition. This was shown by the correlations with intelligence, responsibility and competence. In the social condition, this correlation disappears. Furthermore, for participants with higher robot experience, the robot was seen as more inert, less human-like and less animated. In the social condition, the robot was rated as being more negative (e.g.:inert, foolish, animacy and more ignorant). Similar effects arose for the robot experience. These participants rated the robot as more positive in the social condition. Within the whole body study, similar effects were found for the robot and computer experience. What all these correlations could have in common is that for each type of user, different expectations influence the perception of the robot. Here, the expectations in question are those that people have towards a robotic system, and not towards people that are transported into a robotic system. In some cases, it could be difficult to fulfill those expectations because they are often created by movies, literature and so forth. Younger users seem to have higher expectations of the cognitive abilities of the robot. More computer experienced users seem to have more expectations of the movement capabilities, and tended to rate the robot as foolish. Whereas participants with more robot experience, seemed to have fewer expectations of the social competences of the robot. This shows that the target group interacting with the robot makes a difference in how robots are perceived, a point that should be considered when designing robot behavior.

Chapter 9
Conclusions

The main goal of this thesis was to investigate the effects of the social behavior of robots on human participants. We distinguished between social behavior, that describes only the social aspects but provides no task relevant information, and functional behavior, that does provide task relevant information. Past research did not make this distinction. Because of this, social behavior was evaluated only in the context of functional behavior. This distinction allows for a more detailed investigation of the topic. The effects tested were if social behavior on its own is capable of make robots appear more familiar and lively. It was hypothesized that social behavior influences the arousal of the human participants, and therefore can positively influence task performance in cooperative situations. These studies found that both assumptions can be made, and that social behavior, generated with DBCs, is sufficient to make a robot appear more familiar, more lively and can increase performance in easy tasks, by arousing the human user.

The research encompassed several stages. After reviewing the theoretical background and the effects of robots on humans, a typology of signals and cues was created that applied sign theory to social robots. Out of this typology, the necessary distinction was made between functional and background behavior, and then used to define Dynamic Background Cues. These are unintentional, dynamic features of the robot that provide no task relevant information such as breathing and blinking. By definition these DBCs, are very subtle although it is hypothesized that these small signs can have large effects onto the human observer. Several studies were used to test the first two Hypotheses and to improve DBCs for the final interaction study that was used to proof the third hypothesis.
The first hypothesis stated that increasing human-likeness using DBCs in-

9. CONCLUSIONS

creases the familiarity and the perceived liveliness of a robot in a cooperative task, and was supported in all studies (chapters 6 and 8). Furthermore in each study different effects showed that the participants perceived the robot as more positive, although not the same items show these significances in all studies. This shows that the different DBCs affect different aspects of how the robot is perceived, but that the general perception is influenced positively. In addition, in all studies it was reported that the more social conditions were more pleasant than the non-social conditions. Items that referred to anthropomorphism were rated higher when DBCs were activated, and participants reported feeling less disturbed by the behavior of the robot. It can be said that DBCs fulfill the first Hypothesis in some way or the other whereby the effect is most prominent when more friendly character traits are shown.

During the same online studies, the second Hypothesis was supported. It states that parameterized DBCs make it possible to create different character traits for a social robot. The online studies showed that depending on the parameter of the DBCs, it is possible to achieve this. According to the study, the synced, medium or the social condition create the most friendly character. Because of this, these parameters were used within the interaction study, showing that they generate the same effect in a real robot. These conditions were rated as relatively more calm, not very dominant, social, likable and attentive. For the other conditions, the robot was rated as for example more dominant when using exaggerated lip movements or not breaking up mutual gaze. The parameters for each cue have rich possibilities for generating different traits. The difficulty is in finding the right combination for all DBCs to achieve a consistent character trait where all cues fit together.

Based on the findings of the previous studies, it was possible to investigate the effects of social robots with a friendly character on humans in a cooperative task. This study found evidence that interacting with a robot that uses more Dynamic Background Cues increases the arousal of a person as documented by objective biophysiological measures (compare section 8.2). This result stands in contrast to the third Hypothesis which states that DBCs decrease arousal when interacting with a robot. Nevertheless, social behavior can increase the task performance of easy tasks by encouraging the user. According to the Yerkes-Dodson Law, these correlations are logical, assuming that the situation, the user is in, is a positive one. In addition, when arousal is positive and relatively low, performance is low, too. The robot that behaves more socially seems to generate positive arousal that motivates the user to solve tasks faster. Ashby et al. [5] found that positive arousal is able to increase certain cognitive abilities by elevating dopamine levels, further supporting the findings presented here. In theory, as soon as tasks become too difficult, the additional arousal could become disturbing. It should be investigated further if for more difficult tasks, social behavior should be deactivated to increase task performance.

In order to generate all the complex behavior of robots, a robotic system was created that allows the integration of evaluated DBCs into another functional system, as well as creating portability for the social behavior onto other robots (compare chapter 7). Therefor it was of great importance to create a flexible solution by using standardized interfaces. These were used for communication between all components. In addition, a level of abstraction was used to make it possible to generate all social behaviors autonomously to ease interaction with an attached functional system. During the development of the robotic system, four criteria for social robots were developed that were realized in this system, and which should help in general to design social robots (compare chapter 4.1).

The first two criteria are more user-centered, and postulate first to design the appearance of the robot in such a way that it does not frighten the user, and that it does not evoke expectations of functions that cannot be fulfilled. Second, the movements of the robot have to fit its appearance, and have to fulfill the evoked expectations. The last two criteria are developer-centered. The first states that a robotic system should have high usability and maintainability. The last criteria concerns exchangeability. It states that it is advisable to use standardized interfaces to facilitate compatibility between different components and different robotic hardware. To realize the user-centered criteria, it was decided to use the iCub as a platform and to create the DBCs to generate behavior fitting the appearance of the robot. On the developer-centered side several mechanisms were created that improve exchangeability and usability. For example the `BasicMovementDatabase`, the PiRA-Editor and the PiRA-XML.

During the process of researching this thesis, several ideas for future work emerged which unfortunately could not be further evaluated. First it would be interesting to test the effects of more DBCs on human participants to gather a more extensive set of available cues. Further, it is desirable to produce an easier way to create natural movements instead of configuring them manually.

Within the results of the different studies, some effects occurred that would

9. CONCLUSIONS

be interesting topics for ongoing studies. During all the studies mainly three attributes of the users showed correlations that revealed an influence on how the robot was perceived (chapters 6 and 8). These attributes were age, computer experience and robot experience. Depending on how old or how experienced the users were it seemed that the users had different expectations for the robot. For example, younger people tend to have higher expectations of its cognitive abilities. People with greater robot experience had lower expectations in the robot's social skills. This shows that different types of users have different expectations for and of robots that should be considered when designing a social robot. In addition, it should be taken into account that society is changing over time. In the future, there will be more and more people who use robots and more people who have experience with technical devices. This could also change expectations on robots. It would be interesting to see how these three attributes affect the perception of DBCs.

As described for the third hypothesis, unexpectedly arousal is not decreased but instead increases (section 8.2). At first sight, this stands in conflict with the findings from Kulić et al.'s [87] study who found that more social behavior in a robot decreases anxiety and therewith arousal. After analyzing the two situations more carefully, the differences becomes clear. In this study, general interaction was much more positive than in Kulić et al.'s study, and in addition, positive arousal was generated, instead of decreasing negative arousal. It would be interesting to test if there is a correlation between these two effects, and if the reaction to social behavior switches in some point if the valence of the situation becomes more positive.

Related work on social behavior confounded background and functional behavior, making it difficult to disambiguate whether the source of an effect was functional or background, or even possibly both. This research provides clues that both types of behavior affect the user on different levels. In all studies the background behavior of the robot mainly influenced the social aspects perceived by the user. The perceived cognitive abilities of the robot remained nearly unchanged. In similar studies concerning social robots that used both behaviors in parallel, the perceived cognitive abilities of the robot were often influenced as well. If this is the case, it is possible that background and functional behavior could be decoupled. This could mean that the influence of background behavior is directed towards social aspects and that of functional behavior to the cognitive aspects of the robot. Perhaps there are even more dividable subgroups within the behavior. The typology might help to find more biologically grounded distinctions.

To conclude, Dynamic Background Cues, even though being very subtle, appear to have large effects on how humans perceive a social robot. They can make robots more familiar and lively, are able to generate a variety of characteristics, and can improve performance by arousing the human participant. Although functional behavior is at least as important because it generates the basis for each fruitful Human-Robot Interaction, by using DBCs interactions can be improved and made more pleasant. Especially for users with little experience with robots, social background behavior could make the little difference between acceptance and refusal of robots.

Appendix A

Questionnaires

A.1 Proof of concept

> Liebe Teilnehmerin/ lieber Teilnehmer,
>
> Bitte lesen Sie jede Aussage genau durch und klicken Sie als Antwort die Kategorie an, die Ihre Sichtweise am besten ausdrückt. Denken Sie nicht zu lange nach, sondern orientieren Sie sich an Ihrer ersten Reaktion auf die Frage. Auch wenn Ihnen einmal die Entscheidung schwer fallen sollte, kreuzen Sie trotzdem immer eine Antwort an, und zwar die, welche Ihrer Meinung nach am ehesten zutrifft. Wenn Sie noch Fragen haben, wenden Sie sich bitte jetzt an Ihre/n Versuchsleiter/In.
>
> Wenn alle Fragen geklärt sind, beginnen Sie bitte jetzt mit der Beantwortung!
>
> [Weiter]

Figure A.1: Questionnaire - Proof of concept 1: Introduction to the questionnaire

iCub ...	trifft überhaupt nicht zu	trifft eher nicht zu	weder noch	eher zutreffend	trifft voll und ganz zu
... ist eher zurückhaltend, reserviert.	○	○	○	○	○
... schenkt anderen leicht Vertrauen, glaubt an das Gute im Menschen.	○	○	○	○	○
... ist bequem, neigt zur Faulheit.	○	○	○	○	○
... ist entspannt, lässt sich durch Stress nicht aus der Ruhe bringen.	○	○	○	○	○
... hat nur wenig künstlerisches Interesse.	○	○	○	○	○
... geht aus sich heraus, ist gesellig.	○	○	○	○	○
... neigt dazu, andere zu kritisieren.	○	○	○	○	○
... erledigt Aufgaben gründlich.	○	○	○	○	○
... wird leicht nervös und unsicher.	○	○	○	○	○
... hat eine aktive Vorstellungskraft, ist fantasievoll.	○	○	○	○	○

Seite 2/5

Figure A.2: Questionnaire - Proof of concept 3: Big5 Inventory

A. QUESTIONNAIRES

1. Denken Sie, das iCub Ihnen persönlich eher unähnlich ist?
   eher unähnlich ○ ○ ○ ○ ○ ○ ○ eher ähnlich

2. Wie nah fühlen Sie sich iCub?
   gar nicht nah ○ ○ ○ ○ ○ ○ ○ sehr nah

3. Wie sehr fühlen Sie sich mit iCub verbunden?
   gar nicht verbunden ○ ○ ○ ○ ○ ○ ○ sehr verbunden

4. Denken Sie, dass Sie und iCub viele Gemeinsamkeiten haben?
   gar keine ○ ○ ○ ○ ○ ○ ○ sehr viele

5. Würden Sie sagen, dass Sie und iCub auf einer Wellenlänge liegen?
   gar nicht ○ ○ ○ ○ ○ ○ ○ sehr

6. Würden Sie gerne mit iCub zusammenleben?
   gar nicht ○ ○ ○ ○ ○ ○ ○ sehr

7. Hat Ihnen die Interaktion mit iCub Spaß gemacht?
   gar nicht ○ ○ ○ ○ ○ ○ ○ sehr

8. Fanden Sie die Interaktion mit iCub angenehm?
   gar nicht ○ ○ ○ ○ ○ ○ ○ sehr

9. Würden Sie sich von iCub im Haushalt helfen lassen?
   gar nicht ○ ○ ○ ○ ○ ○ ○ sehr gerne

10. Würden Sie iCub auf Ihre Kinder aufpassen lassen?
    gar nicht ○ ○ ○ ○ ○ ○ ○ sehr gerne

11. Würden Sie iCub Ihre Kreditkarte leihen?
    gar nicht ○ ○ ○ ○ ○ ○ ○ sicher

12. Würden Sie iCub kaufen?
    gar nicht ○ ○ ○ ○ ○ ○ ○ sehr gerne

13. Wie menschlich wirken iCub's Bewegungen auf Sie?
    unmenschlich ○ ○ ○ ○ ○ ○ ○ sehr menschlich

14. Haben Sie das Gefühl, iCub hat Sie verstanden?
    garnicht verstanden ○ ○ ○ ○ ○ ○ ○ hat mich verstanden
    Begründung: [   ]

15. Sind Sie der Meinung, iCub könnte in einer Woche das Gezeigte wiederholen?
    unmöglich ○ ○ ○ ○ ○ ○ ○ ohne Probleme

16. Meinen Sie iCub kann die Übungen nachmachen?
    unmöglich ○ ○ ○ ○ ○ ○ ○ ohne Probleme

17. War iCub ein guter Schüler?
    gar nicht ○ ○ ○ ○ ○ ○ ○ sehr

18. Hat iCub Ihnen zugehört?
    gar nicht ○ ○ ○ ○ ○ ○ ○ sehr aufmerksam

19. Gehen Sie davon aus, das iCub selbstständig gehandelt hat?
    nicht selbstständig ○ ○ ○ ○ ○ ○ ○ komplett selbstständig

20. Schätzen Sie wie alt der Roboter ist!
    ca. [   ] Jahre

21. Ist Ihnen ein Unterschied im Verhalten von iCub aufgefallen, als Sie ihm a) die Lampe und b) die Klingel gezeigt haben?
    Ja ○
    Nein ○
    Begründung: [   ]

Seite 1/5   [Weiter]

Figure A.3: Questionnaire - Proof of concept 2: General questionnaire (23 Items)

A.2 Online Studies

Studie 2/3

Liebe Teilnehmerin/ lieber Teilnehmer,

Im Rahmen dieser Studie werden Ihnen nun 3 Videos gezeigt, die Sie sich aufmerksam ansehen sollen. (Bitte stellen Sie sicher, dass Sie den Sound des Videos hören können) Diese Videos dauern jeweils nur einige Sekunden und zeigen den Roboter iCub.

Nach jedem Video werden Ihnen ein paar Fragen zu dem Video gestellt. Bitte lesen Sie jede Frage genau durch und klicken Sie als Antwort die Kategorie an, die Ihre Sichtweise am besten ausdrückt. Denken Sie nicht zu lange nach, sondern orientieren Sie sich an Ihrer ersten Reaktion auf die Frage. Auch wenn Ihnen einmal die Entscheidung schwer fallen sollte, kreuzen Sie trotzdem immer eine Antwort an, und zwar die, welche Ihrer Meinung nach am ehesten zutrifft. Bei dringenden Fragen kontaktieren Sie mich bitte unter "sgieselm@cor-lab.uni-bielefeld.de" bevor Sie anfangen.

Bitte machen Sie während der Studie keinen Gebrauch von den Navigationstasten Ihres Browsers, andernfalls würden Ihre Eingaben verfälscht. Um den Fragebogen besser lesen zu können empfehlen wir Ihnen in den Vollbildmodus zu wechseln (z.B.: "F11" im Firefox oder "ALT-Enter" im InternetExplorer). Außerdem sollten sie die Studie alleine durchführen, damit sie nicht abgelenkt werden.

Viel Spaß!!!

Figure A.4: Instructions for the online studies

Figure A.5: Video page from the online studies

A. QUESTIONNAIRES

iCub war ...
nervös	○	○	○	○	○	entspannt
aufmerksam	○	○	○	○	○	desinteressiert
ruhig	○	○	○	○	○	aufgedreht/erregt
positiv	○	○	○	○	○	negativ
dominant	○	○	○	○	○	zurückhaltend

Wie menschenähnlich empfanden sie die Bewegungen des Roboters?
Nicht menschenähnlich ○ ○ ○ ○ ○ sehr menschenähnlich
Begründung (optional): [   ]

Die Blickbewegung von iCub waren...
zu schnell	○	○	○	○	○	zu lansam
angenehm	○	○	○	○	○	unangenehm
störend	○	○	○	○	○	haben nicht gestört

Der Blickkontakt vom iCub war...
zu lang ○ ○ ○ ○ ○ zu kurz

[Weiter]

Figure A.6: Questionnaire for the gaze online study

iCub war ...
nervös	○	○	○	○	○	entspannt
aufmerksam	○	○	○	○	○	desinteressiert
ruhig	○	○	○	○	○	aufgedreht/erregt
positiv	○	○	○	○	○	negativ
dominant	○	○	○	○	○	zurückhaltend

Wie menschenähnlich empfanden sie die Bewegungen des Roboters?
Nicht menschenähnlich ○ ○ ○ ○ ○ sehr menschenähnlich
Begründung (optional): [   ]

Die Lippenbewegung von iCub waren...
synchron	○	○	○	○	○	asynchron
zu schnell	○	○	○	○	○	zu lansam
angenehm	○	○	○	○	○	unangenehm
störend	○	○	○	○	○	haben nicht gestört

[Weiter]

Figure A.7: Questionnaire for the lip online study

A.3 Usability Studies

PiRA Editor – Jeder kann Roboter steuern

Herzlich Willkommen bei unserer Studie über den PiRA Editor, ein Anwendungsprogramm mit dem es möglich ist Bewegungsabläufe für Roboter zu Programmieren.

Um die Anwendbarkeit des Programmes zu testen möchten wir Sie bitten die folgenden Aufgaben so schnell wie möglich zu erledigen.

1. Einarbeitung

- Suchen Sie aus der Liste das Gelenk „rightarm_elbow" und wählen Sie es aus. ____min
- Erstellen Sie mit Rechtsklick 5 zusätzliche Ankerpunkte mit verschiedenen Werten.
  (die X-Achse ist die Zeit(Time), die Y-Achse der Gelenkwinkel(Angle)) ____min

- Suchen Sie aus der Liste das Gelenk „leftarm_elbow" und wählen Sie es aus. ____min
- Erstellen Sie wieder 5 Anker, doch diesmal so,
  dass sie an den selben Stellen sind, wie die des rechten Ellenbogengelenkes. ____min

- Verschieben Sie einen Anker des linken Ellenbogengelenks
  auf der Zeitachse auf 500ms und setzen Sie den Winkel auf 60°. ____min
- Verschieben Sie nun alle Anker des rechten Ellenbogengelenks so,
  dass der Anker mit dem höchsten Winkel den Maximalwert erreicht,
  alle Zeitabstände aber erhalten bleiben.
  (Tipp: mit halten von Strg können mehrere Anker ausgewählt werden.) ____min

- Schauen Sie sich ihre selbst generierte Bewegung auf dem Emulator an. ____min

Gesamtzeit zum einarbeiten: ____min

2. Eine Bewegung Imitieren

- Laden Sie die vorbereitete Datei „winken.bee" und schauen sich die Bewegung an. ____min
- Verändern Sie die Einstellungen so,
  dass die Bewegung mit rechts statt mit links durchgeführt wird.
  (Auch die mittleren Körperbewegungen müssen angepasst werden.) ____min

Gesamtzeit zum anpassen: ____min

3. Eine Bewegung Programmieren

- Lassen Sie den Roboter eine Verbeugung machen. Hierbei soll eine Hand vor die Brust gehalten werden, der andere Arm wird abgespreizt. (mittelalterl. „Kratzfuß" Verbeugung) Auf der Körperseite wo sich der Arm vor dem Körper befindet kann der Fuß hierbei ein kein wenig zurückgezogen werden. ____min

Gesamtzeit : ____min

Figure A.8: Exercise sheet for the PiRA-Editor

A. QUESTIONNAIRES

Aufgaben

Herzlich Willkommen bei unserer Vergleichsstudie. Hier lernen Sie Choregaphe kennen, ein Anwendungsprogramm mit dem es möglich ist Bewegungsabläufe für Roboter zu Programmieren.

Um die Anwendbarkeit des Programmes zu testen möchten wir Sie bitten die folgenden Aufgaben so schnell wie möglich zu erledigen.

1. Einarbeitung

- Suchen Sie aus der Liste das Gelenk „rightarm_elbow_roll" und wählen Sie es aus. ____ min
- Erstellen Sie mit Rechtsklick 5 zusätzliche Ankerpunkte mit verschiedenen Werten.
  (die X-Achse ist die Zeit(Time), die Y-Achse der Gelenkwinkel(Angle)) ____ min
- Suchen Sie aus der Liste das Gelenk „leftarm_elbow_roll" und er stellen Sie wieder 5 Anker. ____ min
- Für die nächste Aufgabe müssen die beiden Einstellungen verglichen werden. Wählen Sie daher aus der Liste alle Gelenke oder mindestens die beiden oben erwähnten aus. ____ min
- Verschieben Sie die 5 Punkte des linken Gelenkes so, dass der Roboter die gleichen bewegungen mit beiden Arben macht.
  (Tipp: die Werte müssen an der X-Achse gespiegelt werden, damit die Gelenkeinstellungen identisch sind) ____ min
- Verschieben Sie einen Anker des linken Ellenbogengelenks
  auf der Zeitachse auf 500ms und setzen Sie den Winkel auf -85.
  (Tipp: es gibt hier eine nützliche Funktion „Curve Key Edit") ____ min

Gesamtzeit zum einarbeiten: ____ min

2. Eine Bewegung Imitieren

- Laden Sie die vorbereitete Datei „winken" und schauen sich die Beweung an. ____ min
- Verändern Sie die Einstellungen so,
  dass das Winken mit rechts statt mit links durchgeführt wird.
  (Wählen Sie hierfür am besten ein Gelenk nach dem anderen so wie das entsprechende Gelenk der anderen Körperseite und passen Sie die Anker an, wie Sie es in Teil 1 gelernt haben.) ____ min

Gesamtzeit zum anpassen: ____ min

3. Eine Bewegung Programmieren

- Lassen Sie den Roboter eine Verbeugung machen. Hierbei soll eine Hand vor die Brust gehalten werden, der andere Arm wird abgespreizt.
  (Beachten Sie, dass der Roboter im Simulator „am Oberkörper aufgehangen" ist. Das Vorbeugen verändert daher nicht die Position des Oberkörpers sondern der Beine. Der tatsächliche Roboter würde sich bei selber Anweisung beugen.) ____ min

Gesamtzeit: ____ min

Figure A.9: Exercise sheet for Choregraphe

PiRA Editor – Jeder kann Roboter steuern

Bitte helfen Sie uns das Programm zu verbessern indem Sie uns Ihren Eindruck und ihre Meinung schildern.

Wie haben Sie die Bedienung des Programms empfunden? Als:

leicht	☐	(intuitiv/schnell verständlich)
normal	☐	(mit mehr Übung vermutlich leicht zu bedienen)
schwer	☐	(kompliziert/nicht leicht nachvollziehbar)

Wie empfanden Sie die Anordnung der einzelnen Komponenten (Buttons, Menüpunkte, etc.)?

Als: sehr gut (intuitiv) ☐   normal ☐   nicht gut (schwer zu finden) ☐

Ist die Benennung der einzelnen (Menü-) Punkte treffend?

ja, absolut ☐ ☐ ☐ ☐ ☐ nein, gar nicht

Anmerkungen: _____
_____

Hat Ihnen die Hilfefunktion geholfen?

Ja ☐   Nein ☐, weil: _____

Worauf sollte in der Hilfe zusätzlich/ besser eingegangen werden?
_____

oder:

Ich hab die Hilfefunktion nicht genutzt ☐

Was fanden Sie an der Bedienung des Programms gut: _____
_____
_____
_____ *

Was Fanden Sie an der Bedienung nicht gut: _____
_____
_____
_____ *

Sonstige Anmerkungen: _____
_____
_____
_____ *

*schreiben Sie bei bedarf auch auf der Rückseite weiter.

Vielen Dank für ihre Hinweise!

Figure A.10: Questionnair used to rate the usability of the editors

A. QUESTIONNAIRES

PiRA Editor – Jeder kann Roboter steuern

Vielen dank, dass Sie an underer Studie zur Verbesserung unseres Programmes teilgenommen haben. Abschließend benötigen wir noch ein paar persönliche Angaben für die Statistische Erhebung. Wir verschern Ihnen, dass mit ihren Angaben absolut vertraulich umgegangen wird und sie allein zur statistischen Auswertung dienen.

Alter: _____   **Geschlecht:** m ☐   w ☐
Muttersprache/n: _____

Englische Sprachkenntnisse: nein ☐   kaum ☐   schulenglisch ☐   ja ☐   sehr gut ☐

Bildungsanschluss:
(wenn Sie mehrere Abschlüsse haben, genügt der aktuellste)

kein Abschluss ☐   Haupt-/Sonderschulabschluss ☐   Mittlere Reife/Realschulabschlss ☐

Abitur/Fachabitur/Hochschulreife ☐   Berufsausbildung ☐ (Berufsfeld:_____)

Bachelorabschluss ☐   Master/ Diplom ☐   Doktor ☐

Informatik und Computerkenntnisse bisher:
(aus der ersten Spalte bitte eins ankreuzen.
Die Übrigen Möglichkeiten sind dann zur genaueren Bestimmung.
Es kann auch Zeilenübergreifend angekreuzt werden.)

keine ☐   - nur mit wenien Programmen ☐   spezielle ☐
einige ☐   - mit verschiedennen Programmen ☐   allgemeine ☐
viele ☐   - mit eigenen Einstellungen/Veränderungen ☐   geringe programmiererfahrung ☐
sehr viele ☐   - mit verschiedennen Betriebssystemen ☐   Programmiererfahrung ☐

Figure A.11: Questionnair for personal data

A.4 Whole Body Study

iCub war ...
...unattraktiv	○	○	○	○	○	...attraktiv
...emotional	○	○	○	○	○	...gefühlskalt
...unsozial	○	○	○	○	○	...sozial
...dumm	○	○	○	○	○	...intelligent
...entspannt	○	○	○	○	○	...nervös
...inkompetent	○	○	○	○	○	...kompetent
...ruhig	○	○	○	○	○	...aufgedreht/erregt
...passiv	○	○	○	○	○	...aktiv
...zurückhaltend	○	○	○	○	○	...dominant
...positiv	○	○	○	○	○	...negativ
...unaufmerksam	○	○	○	○	○	...aufmerksam
...unglaubwürdig	○	○	○	○	○	...glaubwürdig
...natürlich	○	○	○	○	○	...künstlich
...unsympathisch	○	○	○	○	○	...sympathisch

Die Bewegungen von iCub ...
...waren steif	○	○	○	○	○	...waren elegant
...waren angenehm	○	○	○	○	○	...waren unangenehm
...haben nicht gestört	○	○	○	○	○	...haben gestört

[Weiter]

Figure A.12: First part of the questionnaire used to rate the robot

Meinen Sie der Roboter könnte Ihr Freund werden?
Eher nicht ○ ○ ○ ○ ○ Sehr gerne

Wie nah fühlen Sie sich dem Roboter?
Gar nicht nah ○ ○ ○ ○ ○ Sehr nah

Mögen Sie den Roboter?
Mag ihn gar nicht ○ ○ ○ ○ ○ Mag ihn sehr

Können Sie das Verhalten des Roboters einschätzen?
überhaupt nicht ○ ○ ○ ○ ○ Sehr gut

Würden Sie gerne mit dem realen Roboter interagieren?
Lieber nicht ○ ○ ○ ○ ○ Sehr gern

Wie lebendig wirkt der Roboter auf Sie?
Gar nicht lebendig ○ ○ ○ ○ ○ Sehr lebendig

Wie menschenähnlich wirkt der Roboter auf Sie?
Gar nicht menschenähnlich ○ ○ ○ ○ ○ Sehr menschenähnlich
Begründung (optional): [   ]

Vertrauen Sie den Informationen des Roboters?
Gar nicht ○ ○ ○ ○ ○ Sehr

Meinen Sie der Roboter kann selbständig agieren?
Kann er nicht ○ ○ ○ ○ ○ Kann er

[Weiter]

Figure A.13: Second part of the questionnaire used to rate the robot

A. QUESTIONNAIRES

A.5 Interaction Study

GODSPEED I: ANTHROPOMORPHISM
Please rate your impression of the robot on these scales:
以下のスケールに基づいてこのロボットの印象を評価してください。

	1	2	3	4	5	
Fake 偽物のような	1	2	3	4	5	Natural 自然な
Machinelike 機械的	1	2	3	4	5	Humanlike 人間的
Unconscious 意識を持たない	1	2	3	4	5	Conscious 意識を持っている
Artificial 人工的	1	2	3	4	5	Lifelike 生物的
Moving rigidly ぎこちない動き	1	2	3	4	5	Moving elegantly 洗練された動き

GODSPEED II: ANIMACY
Please rate your impression of the robot on these scales:
以下のスケールに基づいてこのロボットの印象を評価してください。

	1	2	3	4	5	
Dead 死んでいる	1	2	3	4	5	Alive 生きている
Stagnant 活気のない	1	2	3	4	5	Lively 生き生きとした
Mechanical 機械的な	1	2	3	4	5	Organic 有機的な
Artificial 人工的な	1	2	3	4	5	Lifelike 生物的な
Inert 不活発な	1	2	3	4	5	Interactive 対話的な
Apathetic 無関心な	1	2	3	4	5	Responsive 反応のある

GODSPEED III: LIKEABILITY
Please rate your impression of the robot on these scales:
以下のスケールに基づいてこのロボットの印象を評価してください。

	1	2	3	4	5	
Dislike 嫌い	1	2	3	4	5	Like 好き
Unfriendly 親しみにくい	1	2	3	4	5	Friendly 親しみやすい
Unkind 不親切な	1	2	3	4	5	Kind 親切な
Unpleasant 不愉快な	1	2	3	4	5	Pleasant 愉快な
Awful ひどい	1	2	3	4	5	Nice 良い

GODSPEED IV: PERCEIVED INTELLIGENCE
Please rate your impression of the robot on these scales:
以下のスケールに基づいてこのロボットの印象を評価してください。

	1	2	3	4	5	
Incompetent 無能な	1	2	3	4	5	Competent 有能な
Ignorant 無知な	1	2	3	4	5	Knowledgeable 物知りな
Irresponsible 無責任な	1	2	3	4	5	Responsible 責任のある
Unintelligent 知的でない	1	2	3	4	5	Intelligent 知的な
Foolish 愚かな	1	2	3	4	5	Sensible 賢明な

GODSPEED V: PERCEIVED SAFETY
Please rate your emotional state on these scales:
以下のスケールに基づいてあなたの心の状態を評価してください。

	1	2	3	4	5	
Anxious 不安な	1	2	3	4	5	Relaxed 落ち着いた
Agitated 動揺している	1	2	3	4	5	Calm 冷静な
Quiescent 平穏な	1	2	3	4	5	Surprised 驚いた

Figure A.14: Goodspeed Questionnaire created by Bartneck et al. Image taken from [6]

Item (German)	Item (English)	Type
Wie einfach/schwer waren die Aufgaben?	How easy/difficult were the tasks?	5 Point scale
Welche Aufgabe fanden Sie am schwersten?	Which task was the most difficult?	Text field
Welche Aufgabe fanden Sie am leichtesten?	Which task was the easiest?	Text field
Begründung?	Explaination?	Text field
Zu welchem Maßhülten Sie sich gestresst?	To what extend did you feel stressed?	5 Point scale
Haben Sie sich unwohl gefühlt?	Did you feel uncomfortable?	5 Point scale
Hat Ihnen die Interaktion Spaßgemacht?	Was it fun to interact with the robot?	5 Point scale
Welche Aufgabe hat Ihnen am meisten Spaßgemacht?	Which task was most fun?	Text field
Welche Aufgabe hat Ihnen am wenigsten Spaßgemacht?	Which task was fewest fun?	Text field
Begründung?	Explaination?	Text field
Wären Sie bereit, die Interaktion zu wiederholen?	How willing would you be to repeat the interaction?	5 Point scale
Wie interessiert waren Sie?	How interested were you?	5 Point scale
Wie aufmerksam waren Sie?	How attentive were you?	5 Point scale
Wie glücklich waren Sie?	How happy were you?	5 Point scale
Wie zufrieden waren Sie?	How satisfied were you?	5 Point scale
Vertrauen Sie den Aussagen des Roboters?	Do you trust the information given by the robot?	5 Point scale
Würden Sie gerne öfter mit dem Roboter interagieren?	Would you like to interact with the robot more often?	5 Point scale
Konnten Sie das Verhalten des Roboters einschätzen?	Are you able to appraise the behavior of the robot?	5 Point scale
Meinen Sie, der Roboter kann selbstständig agieren?	Do you think the robot can act on its own?	5 Point scale

Table A.1: List of items used in the interaction Study. The original German items as well as the English translations are shown.

Appendix B

Tables

B.1 Proof of concept

Question (translated from German to English)	Mdn-non	Mdn-social	U	p	r
1. iCub is diffident, reserved.	4	4	48	.57	0.124
2. iCub trusts others easily and believes in the good in man	3.5	4	51	.769	0.064
3. iCub is ease and tends to laziness	2.5	3	50.5	.736	0.074
4. iCub is relaxed and is not flappable by stress	4	4	37	.178	0.294
5. iCub is not interested in arts	3	3	53	.875	0.034
6. iCub is convivial	2	2	50.5	.738	0.073
7. iCub tends to criticize others	2	2	55	1	0
8. iCub finishes tasks thoughtfully	4	3	45	.472	0.157
9. iCub becomes nervous and uncertain very easy	2	3	41.5	.321	0.216
10. iCub has a vivid imagination	2	3	30.5	.066	0.401

Table B.1: Big5 - Single items (Mann-Whitney Test). Item values range from 1 to 5

Combined Big5 scores	M-non	SD-non	M-social	SD-social	t	df	p	d
Extraversion (1 and 6)	6.3	1.16	6.091	1.136	0.684	19	0.681	0.219
Agreeableness (2 and 7)	5.5	1.509	5.455	1.368	0.072	19	0.943	0.031
Conscientiousness (3 and 8)	6.4	0.699	6.272	0.905	0.358	19	0.724	0.158
Neuroticism (4 and 9)	6	1.054	6.909	1.044	-1.983	19	0.062	0.866
Openness (5 and 10)	5.5	1.179	6.455	0.687	-2.238	19	0.033*	0.989

Table B.2: Big5 - Combined score (T-Test)

B. TABLES

Question (translated from German to English)	Mdn-non	Mdn-social	U	p	r
1. Do you think iCub differs from you on a personal level?	2.5	1	45	.468	0.158
2. How close do you feel to the robot?	3	3	53,5	.912	0.024
3. How much do you feel connected to the iCub?	2	2	46	.491	0.150
4. Do you think you and iCub have much in common?	2.5	2	53	.881	0.032
5. Do you think you and iCub are tuned on the same wavelength?	2	2	50.5	.741	0.072
6. Do you want to live together with the iCub?	3	3	54.5	.971	0.008
7. Did you had fun during the interaction with iCub?	5	4	49	.667	0.094
8. Do you rate the interaction with iCub as pleasant?	4	4	50.5	.746	0.071
9. Would you let iCub help you in the household?	4	6	51	.771	0.063
10. Would you let iCub take care of your children?	2.5	2	35	.143	0.319
11. Would you lend iCub your credit card?	1	1	52	.813	0.051
12. Would you like to buy iCub	4	5	38.5	.237	0.258
13. How human-like are the movements of iCub	3	5	20.5	.013*	0.542
14. Do you have the feeling iCub understood what you taught the robot?	3	3	49	.667	0.094
15. Do you have the feeling iCub could repeat the taught tasks after a week?	3	3	52	.829	0.047
16. Do you think iCub can repeat the shown tasks?	3.5	3	54	.943	0.016
17. Was the iCub a good student?	3.5	2	38.5	.227	0.263
18. Did iCub listen to you?	3.5	2	41.5	.333	0.211
19. Do you think iCub acted by its own?	3	5	43	.389	0.188

Table B.3: General questionnaire (Mann-Whitney Test). Item values range from 1 to 7

B.2 Online Study: Gazing

Items	Mdn -Inf	Mdn -Med	Mdn -Ex	Inf-Med			Inf-Ex			Med-Ex		
				Z	p	r	Z	p	r	Z	p	r
nervous/relaxed	5	5	2	0.4	.687	0.07	3.6	<.001*	0.69	3.68	<.001*	0.7
attentive/disinterested	2	2	3	No Friedmann Sig. ($\chi^2(2) = 1.407, p = .495$)								
calm/aroused	2	2	5	1.2	.231	0.22	4.17	<.001*	0.78	3.92	<.001*	0.74
positive/negative	3	2	3.5	No Friedmann Sig. ($\chi^2(2) = 2.96, p = .228$)								
dominant/submissive	3	5	4	2.83	.005*	0.53	1.96	.05	0.38	2.28	.23	0.44
not human/very human	4	5	4	2.7	.007*	0.5	0.99	.323	0.18	2.12	.034	0.39
too long/too short	1	4	4	4.21	<.001*	0.8	4.51	<.001*	0.84	2.42	<.001*	0.46
too fast/too slow	4	4	2	1.6	.11	0.31	3.99	<.001*	0.77	3.8	<.001*	0.73
pleasant/unpleasant	4	2	4	3.31	.001*	0.63	0.37	0.711	0.07	3.48	<.001*	0.65
disturbing/not disturbing	4	5	4	2.48	.013*	0.49	0.32	.752	0.06	2.88	.004*	0.55

Table B.4: Questionnair results (Wilcoxon Signed-ranks Test) for the online gaze study. Item values range from 1 to 6

B.3 Phoneme/Viseme mapping

MARY	Viseme	iCub
p	P	closed
b	P	closed
m	P	closed
?	P	closed
n	N	closed
@	N	closed
l	N	closed
n	N	closed
=6	N	closed
6	N	closed
A	N	closed
{	N	closed
AI	N	closed
s	S	closed
z	S	closed
h	R	closed
r	R	closed
r=	R	closed
R	R	closed
o:	O	closed
o	O	closed
o~	O	closed
O	O	closed
OI	O	closed
u:	U	closed
u	U	closed
U	U	closed
f	U	closed
v	U	closed
w	U	closed
pf	U	closed
y:	Y	closed
y	Y	closed
Y	Y	closed
2:	Y	closed
2	Y	closed
9	Y	closed
9~	Y	closed
oY	Y	closed
OY	Y	closed
a~	Y	closed

MARY	Viseme	iCub
t	T	open
d	T	open
k	T	open
g	T	open
D	T	open
T	T	open
S	Z	open
Z	Z	open
tS	Z	open
ts	Z	open
dZ	Z	open
j	C	open
C	C	open
x	C	open
i:	E	open
i	E	open
I	E	open
e:	E	open
e	E	open
E:	E	open
E	E	open
a:	E	open
a	E	open
aI	E	open
aU	E	open
EI	E	open
e~	E	open
V	E	open
@U	E	open

Table B.5: List of all necessary phonemes for English and German speech synsthesis. Viseme labels were taken from [4]

B.4 Online Study: Lips

Items	Mdn -Inf	Mdn -Med	Mdn -Ex	noLips-Synced Z	p	r	noLips-Ex Z	p	r	Synced-Ex Z	p	r
nervous/relaxed	4	4	4	\multicolumn{9}{c}{No Friedmann Sig. ($\chi^2(2) = 5.828, p = .054$)}								
attentive/disinterested	4	3	3	3	.003*	0.52	1	.039	0.17	2.07	.317	0.36
calm/aroused	3	3	3	\multicolumn{9}{c}{No Friedmann Sig. ($\chi^2(2) = 1.820, p = .403$)}								
positive/negative	3	3	3	3.5	<.001*	0.61	2.13	.033	0.37	1.88	.061	0.33
dominant/submissive	4	4	4	2.04	.041	0.36	2.45	.014*	0.43	0.53	.594	0.09
not human/very human	1	3	3	4.6	<.001*	0.83	3.78	<.001*	0.68	2.3	.021	0.41
synchronous/asynchronous	6	2	3	4.32	<.001*	0.85	3.74	<.001*	0.73	1.95	.051	0.35
too fast/too slow	6	3	3	3.86	<.001*	0.74	4.14	<.001*	0.8	1.91	.056	0.34
pleasant/unpleasant	5	3	3	4.31	<.001*	0.83	3.47	.001*	0.67	3.05	.002*	0.55
disturbing/not disturbing	3	5	5	3.52	<.001*	0.64	2.39	.017	0.44	2.49	.013*	0.44

Table B.6: Questionnair results (Wilcoxon Signed-ranks Test) for the online lips study. Item values range from 1 to 6

B.5 Online Study: Whole Body

Items	Mdn-non-social	Mdn-social	Z	p	r
unattractive/attractive	4	4	2.62	.009*	0.45
emotional/callously	5	5	4.1	<.001*	0.7
unsocial/social	4	5	3.01	.003*	0.52
dense/intelligent	4	4	0.86	.391	0.15
relaxed/nervous	3	2	1.37	.17	0.24
incompetent/competent	4	5	2.87	.004*	0.49
calm/aroused	2	2	1.81	.07	0.31
passive/active	4	5	3.74	<.001*	0.64
submissive/dominant	3	4	0.32	.747	0.06
positive/negative	3	3	2.76	.006*	0.47
disinterested/attentive	4	5	2.82	.005*	0.48
unreliable/reliable	4	5	1.42	.156	0.24
natural/artificial	5	3	4.25	<.001*	0.73
unlikeable/likeable	4	5	2.65	.008*	0.45
stiff/elegant	2	4.5	4.06	<.001*	0.70
pleasant/unpleasant	4	2	3.46	.001*	0.59
disturbing/not disturbing	3	2	2.66	.008*	0.46
Do you think the robot could become a friend of yours?	2	3	3.27	.001*	0.56
How close do you feel to the robot?	2	3	3	.003*	0.51
Do you like the robot?	4	4.5	2.77	.006*	0.48
Are you able to estimate the behavior of the robot?	3	4	2.31	.21	0.40
Would you like to interact with the real robot?	4	5	1.43	.154	0.24
How vivid did the robot appear to you?	2	4	4.29	<.001*	0.74
How human-like did the robot appear to you?	3	4	4.14	<.001*	0.71
Do you trust the information given by the robot?	4	5	2.01	.045*	0.34
Do you think the robot can act on its own?	2.5	4	2.67	.007*	0.46

Table B.7: Questionnaire results (Wilcoxon Signed-ranks Test) for the online whole body study. Item values range from 1 to 6

B.6 Interaction Study: iCub conversation

Speech (in order of execution)	non-social movement	Social movement	Gaze
First phase			
"Hello."	-	Lift hands to the side slightly	-
"My name is iCub."	Right hand to chest	FB	-
"Thank you for participation in this study."	Right hand points to participant	FB	-
Second phase			
"In the following I will set 6 little task to you, that should be solved as good as possible."	-	-	-
"I will not repeat the instructions."	-	-	-
"Please execute the task as you understood it."	-	Open hands orientated to the participant	-
"And remember not to move the arm with the sensors during the whole study."	Right hand point to sensors	FB	to sensors
Building phase			
"The next task concerns building towers."	-	-	-
"Please build these within the area in front of you and only with the free hand."	Right hand point to center area	FB	to center area
"Do you see the bricks to your right?"	Left hand point to bricks	FB	to bricks
"Please take ten bricks and build a tower that is as stable as possible." (wait)	Move arm to front and hold palms towards each other	FB	to center area after instruction
"Good."	-	-	-
"Please take ten bricks and build a tower that is as small as possible. Thereby no brick is allowed to lay beside th other" (wait)	Show a lower height. Palms down	FB	to center area after instruction
"Good."	Nodding	Nodding	-
"Take four bricks and build the highest tower possible." (wait)	Show a higher height. Palms down	FB	to center area after instruction
"Very well."	Nodding	Nodding	-
"Now we finished the building task. Please put back the bricks."	-	Lift hands to side. Pals to the front	-
Cognitive phase			
"Within this tasks your cognitive ability is needed"	Right hand point to partivipant	FB	-
"Please remember the following items and recall them as good as possible"	-	Lift right hand. Palm to front	-
"Repeat the following three items."	Show three	FB	-
"House, key, baby buggy." (wait)	-	-	-
"OK."	-	-	-
"Repeat the following five items."	Show three	FB	-
"Flower, wheel, telephone, locker, television." (wait)	-	-	-
"Good."	Nodding	Nodding	-
"Repeat the following seven items."	Lift hands to the side. Palms to front	FB	-
"Bag, spoon, trousers, forest, garbage can, street, bed." (wait)	-	-	-
"Thats it."	-	Lift right hand. Palms front	-
Fifth phase			
"Thereby all tasks are solved."	Lift hands to both sides	FB	-
"Thank you again for your participation."	-	-	-
"A colleague of mine will explain everything else to you."	Left hand point to human experimenter	FB	Gaze to human experimenter

Table B.8: Conversation executed by the iCub during the interaction study (Translated from German to English). FB = Same as functional behavior but with natural movement; (Wait) = Robot waits until task is completed

B. TABLES

B.7 Interaction Study: Statistics

Item	M-non	SD-non	M-social	SD-social	t	df	p	d
Negative Affect								
afraid	0.09	0.596	-0.13	0.947	0.914	45	.366	0.278
scared	-0.04	0.562	-0.13	0.850	0.386	45	.701	0.125
nervous	-0.04	0.976	-0.04	0.806	-0.007	45	.994	0
jittery	-0.09	0.596	0.04	1.042	-0.516	45	.608	0.153
irritable	-0.3	0.703	-0.46	0.779	0.71	45	.481	0.216
hostile	-0.13	0.344	-0.08	0.282	-0.514	45	.610	0.159
guilty	0.09	0.417	-0.17	0.637	1.607	45	.115	0.32
ashamed	0.09	0.596	0.13	0.68	-0.204	45	.840	0.063
upset	-0.09	0.515	-0.21	0.588	0.752	45	.456	0.217
distressed	-0.17	0.65	-0.13	0.68	-0.252	45	.802	0.06
Positive Affect								
active	0.09	0.733	-0,13	0.855	0.914	45	.366	0.276
alert	-0.43	0.728	-0.04	1.16	-1.384	45	.173	0.403
attentive	-0.17	0.778	-0.25	0.608	0.375	45	.71	0.115
determined	-0.61	0.941	-0.08	0.974	-1.879	45	.067	0.553
enthusiastic	0.22	1.126	0.38	0.77	-0.562	45	.577	0.166
excited	0.09	1.164	0.21	0.833	-0.412	45	.682	0.119
inspired	0.00	1.414	-0.29	0.999	0.819	45	.417	0.237
interested	0	0.739	-0.29	0.751	1.342	45	.186	0.389
proud	-0.26	0.915	0.25	0.676	-2.183	45	.034*	0.634
strong	-0.52	0.994	-0.17	0.702	-1.419	45	.163	0.407
Fear								
afraid	See above							
scared	See above							
frightened	-0.09	0.996	-0.17	0.917	0.286	45	.776	0.084
nervous	See above							
jittery	See above							
shaky	-0.13	0.815	0.17	1.007	-1.109	45	.273	0.327
Joviality								
happy	-0.26	0.752	-0.08	0.83	-0.768	45	.447	0.227
joyful	-0.09	0.949	-0.08	0.83	-0.014	45	.989	0.011
delighted	-0.09	1.125	-0.04	0.751	-0.163	45	.871	0.052
cheerful	-0.13	0.869	0.08	0.584	-0.994	45	.326	0.284
excited	See above							
enthusiastic	See above							
lively	-0.52	1.275	0.04	0.859	-1.784	45	.081	0.515
energetic	-0.09	0.9	0.04	0.751	-0.188	45	.852	0.06

Table B.9: PANAS test for relative self-assessment(T-Test). Item values range from 1 to 5

Item	Mdn-non	Mdn-social	U	p	r
GODSPEED I: Anthropomorphism					
fake/natural	3	3	189.5	.054	0.282
machinelike/humanlike	2	3	181.5	.036*	0.306
unconscious/conscious	3	3	230.5	.313	0.147
artificial/lifelike	3	3	195	.071	0.263
moving rigidly/moving elegantly	3	3	258	.686	0.059
GODSPEED II: Animacy					
dead/alive	4	4	262.5	.755	0.046
stagnant/lively	4	4	191	.056	0.28
mechanical/organic	2	2	271.5	.187	0.193
artificial/lifelike	3	3	195	.071	0.263
inert/interactive	4	4	241.5	.431	0.115
apathetic/responsible	4	4	244.5	.487	0.101
GODSPEED III: Likeability					
dislike/like	4	5	225.5	.237	0.173
unfriendly/friendly	5	5	244	.444	0.112
unkind/kind	4	4	253.5	.603	0.076
unpleasant/pleasant	4	4.5	188.5	.04*	0.3
awful/nice	4	4	216.5	.174	0.198
GODSPEED IV: Perceived Intelligence					
incompetent/competent	4	4	249.5	.531	0.091
ignorant/knowledgeable	3	4	235.5	.359	0.134
irresponsible/responsible	3	3.5	260	.717	0.053
unintelligent/intelligent	4	4	234.5	.347	0.137
foolish/sensible	4	4	241.5	.440	0.113

Table B.10: Godspeed questionnaire (Mann-Whitney Test). Item values range from 1 to 5

Question (translated from German to English)	Mdn-non	Mdn-social	U	p	r
1. iCub is diffident, reserved.	4	3	217.5	.194	0.189
2. iCub trusts others easily and believes in the good in man	3	3	238	.387	0.126
3. iCub is ease and tends to laziness	3	2	237	.385	0.127
4. iCub is relaxed and is not flappable by stress	4	4	262.5	.751	0.046
5. iCub is not interested in arts	3	3	207	.126	0.223
6. iCub is convivial	2	3	198	.083	0.253
7. iCub tends to criticize others	1	2	256.5	.652	0.066
8. iCub finishes tasks thoughtfully	5	5	245.5	.471	0.105
9. iCub becomes nervous and uncertain very easy	1	1	243	.384	0.127
10. iCub has a vivid imagination	3	2	211	.151	0.209

Table B.11: Big5 - Single items (Mann-Whitney Test). Item values range from 1 to 5

Combined Big5 scores	M-non	SD-non	M-social	SD-social	t	df	p	d
Extraversion (1 and 6)	5	1.931	5.92	1.666	-1.745	45	.088	0.51
Agreeableness (2 and 7)	7.17	1.302	7.21	1.285	-0.091	45	.928	0.03
Neuroticism (4 and 9)	8.52	1.377	8.88	1.035	-0.997	45	.324	0.296
Openness (5 and 10)	5.83	1.23	4.83	1.903	2.113	45	.04*	0.624
Conscientiousness (3 and 8)	7.87	1.792	8.38	1.313	-1.107	45	.274	0.325

Table B.12: Big5 - Combined score(T-Test)

B. TABLES

Item	Mdn-non	Mdn-social	U	p	r
Negative Affect					
afraid	1	1	262	.577	0.081
scared	1	1	263.5	.53	0.092
nervous	1	1	276	1	0
jittery	1	1	270	.867	0.024
irritable	1	1	243.5	.196	0.189
hostile	1	1	265	.58	0.08
guilty	1	1	252	.144	0.213
ashamed	1	1	251	.357	0.134
upset	1	1	251	.281	0.157
distressed	1	1	268	.812	0.035
Positive Affect					
active	3	4	185	.04*	0.299
alert	3	4	166.5	.016*	0.352
attentive	3	4	217.5	.19	0.191
determined	4	4	223	.233	0.174
enthusiastic	2	2.5	227	.28	0.157
excited	1	2	213	.156	0.207
inspired	2	3	258.5	.697	0.057
interested	3	4	224	.249	0.168
proud	1	2	212	.136	0.218
strong	2	3	217.5	.201	0.187
Fear					
afraid	See above				
scared	See above				
frightened	1	1	252	.378	0.129
nervous	See above				
jittery	See above				
shaky	1	1	243	.379	0.128
Joviality					
happy	2	2	210.5	.145	0.213
joyful	2	3	217.5	.195	0.189
delighted	2	3	206.5	.125	0.224
cheerful	2	2	233	.331	0.142
excited	See above				
enthusiastic	See above				
lively	2	3	215.5	.179	0.196
energetic	3	3	244.5	.488	0.101

Table B.13: PANAS Test for the robot (Mann-Whitney Test). Item values range from 1 to 5

Item	M-non	SD-non	M-social	SD-social	t	df	p	d
Negative Affect	11.7	3.066	11.42	2.062	0.367	45	.715	0.107
Positive Affect	26.65	7.901	30.92	6.554	-2.018	45	.050*	0.588
Difference Pos/Neg	14.96	8.626	19.5	6.108	-2.091	45	.042*	0.607
Fear	7.39	2.251	7.08	1.666	0.535	45	.595	0.157
Joviality	17.52	7.179	20.33	6.638	-1.395	45	.17	0.406

Table B.14: PANAS scores for the robot(T-Test)

Item	Mdn-non	Mdn-social	U	p	r
How easy/difficult were the tasks?	2	2	271.5	.919	0.015
To what extend did you feel stressed?	2	2	270.5	.899	0.019
Did you feel uncomfortable?	2	2	257.5	.675	0.061
Was it fun to interact with the robot?	4	4.5	192	.048*	0.288
How willing would you be to repeat the interaction?	5	5	250.5	.497	0.099
How interested were you?	4	5	230	.272	0.16
How attentive were you?	4	4.5	221	.207	0.184
How happy were you?	3	4	169	.011*	0.37
How satisfied were you?	3	4	214	.157	0.207
Do you trust the information given by the robot?	3	4	205	.116	0.23
Would you like to interact with the robot more often?	4	5	211	.145	0.213
Are you able to appraise the behavior of the robot?	2	3.5	225.5	.256	0.166
Do you think the robot can act on its own?	2	3	222.5	.241	0.171

Table B.15: Functional questionnaire (Mann-Whitney Test). Item values range from 1 to 5

Appendix C

Images

Visem	Visual Representation	Viseme	Visual Representation
P		C	
T		E	
N		A	
M		O	
F		U	
S		Q	
Z		Y	
R			

Figure C.1: Corresponding visemes and lip postures taken from [4]

C. IMAGES

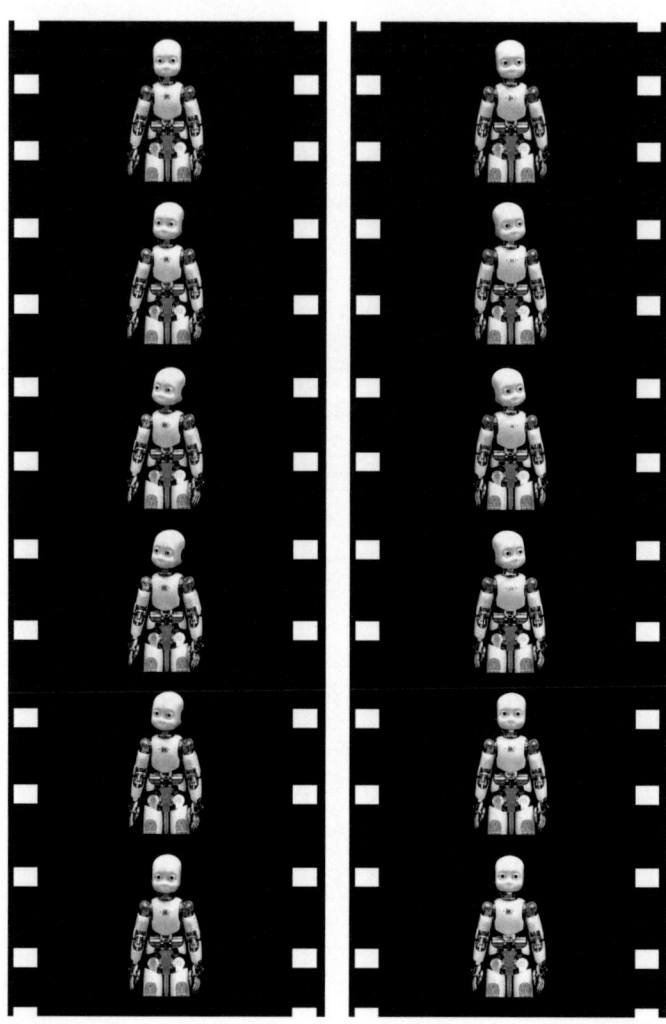

Figure C.2: Idle movements: Looking around as performed by iCub

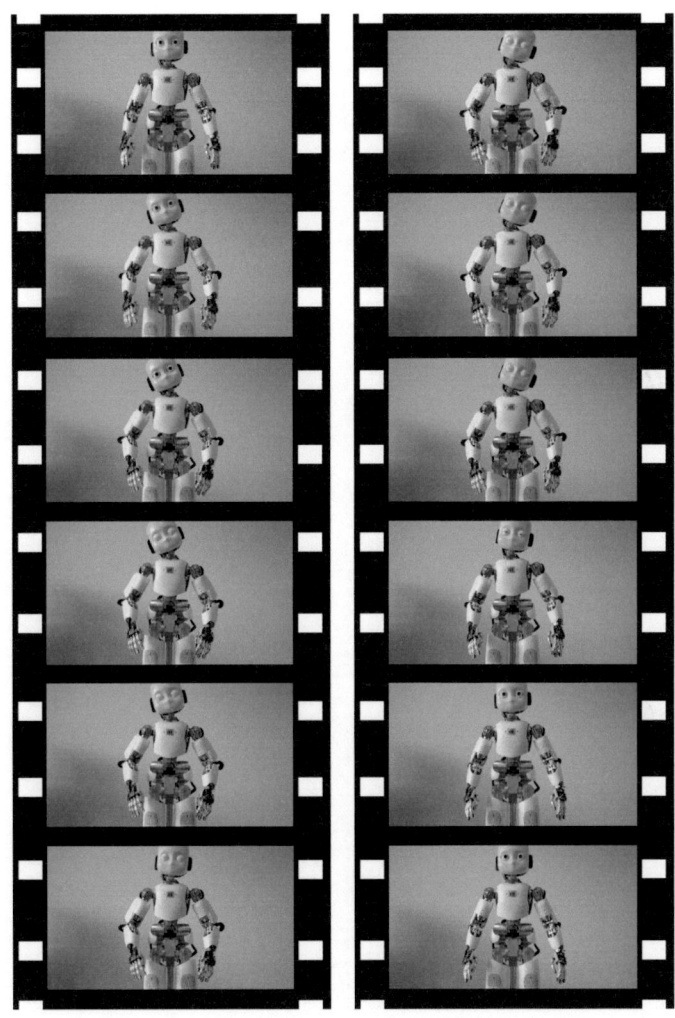

Figure C.3: Idle movements: A stretchung motion performed by iCub

C. IMAGES

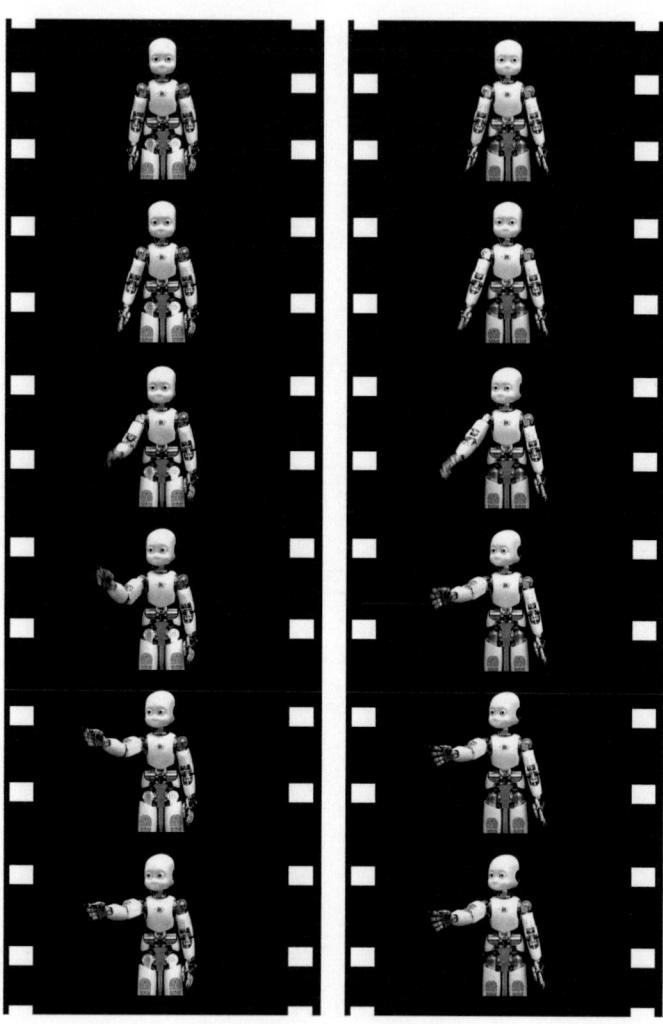

Figure C.4: Pointing gesture: Performed with DBCs (left) and without (right)

Appendix D

UML Diagrams

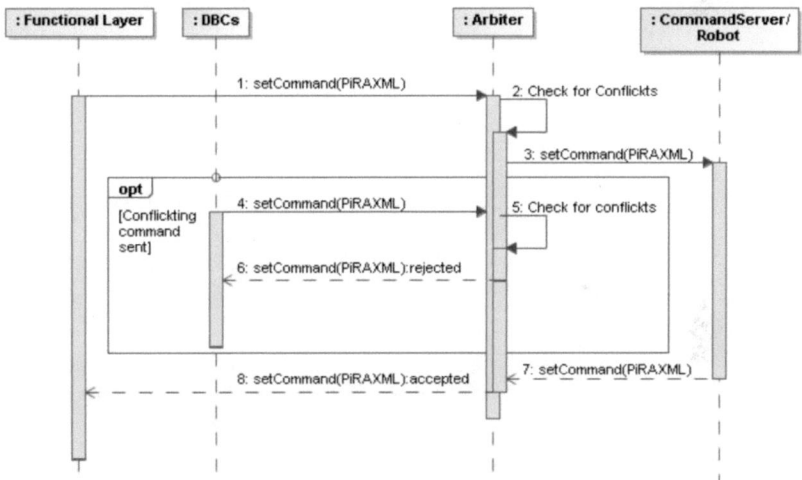

Figure D.1: Sequence Diagram: External command. While an animation with external origin is executed, a DBC tries to execute a conflicting command to the robot itself and gets rejected.

D. UML DIAGRAMS

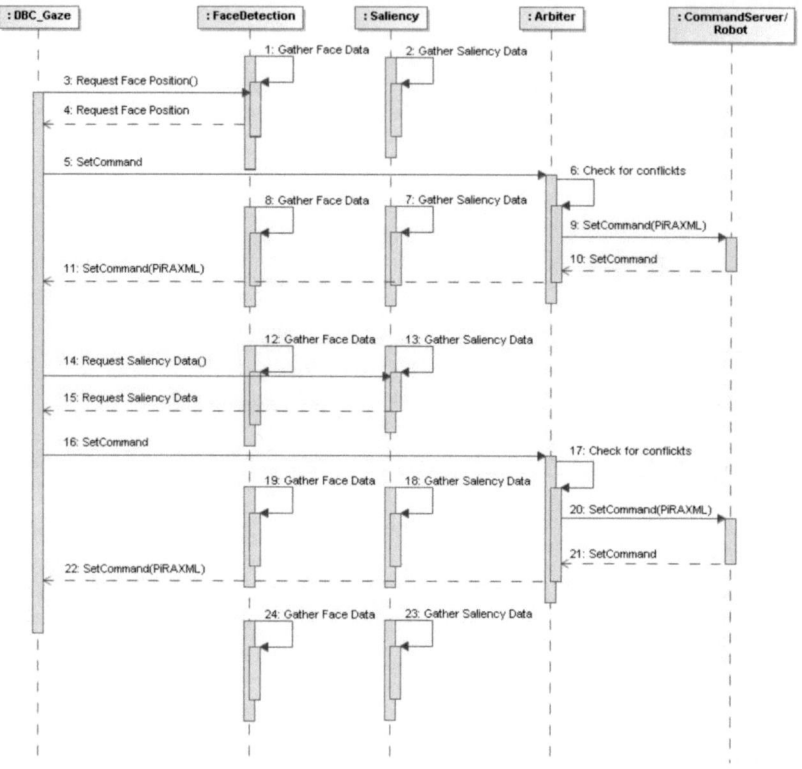

Figure D.2: Sequence Diagram: Gaze DBC. A gaze to a detected face followed by a gaze to a salient spot are performed.

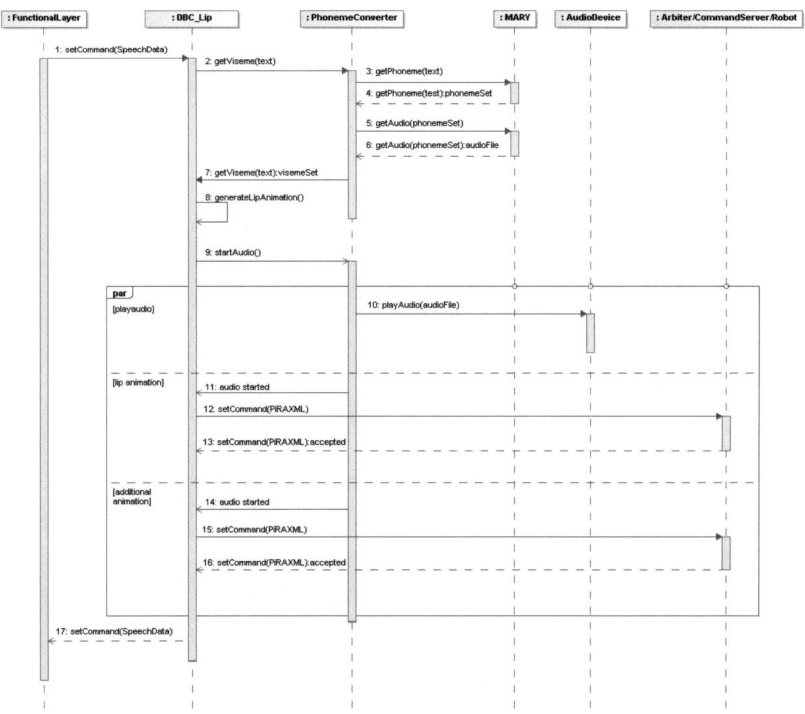

Figure D.3: Sequence Diagram: Speech command. After converting the input text, the audio file, lip movements and additional movements are carried out in parallel

D. UML DIAGRAMS

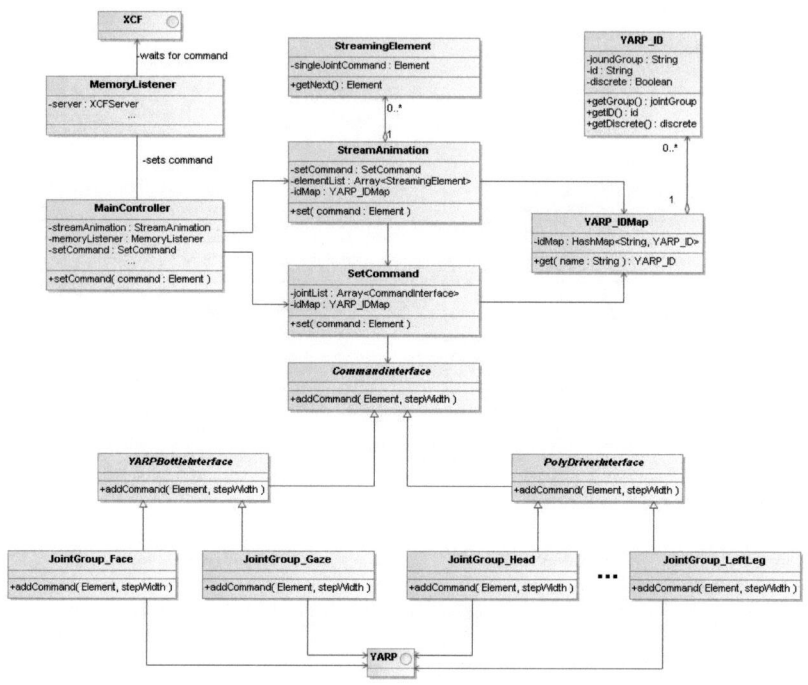

Figure D.4: Class Diagram: `CommandServer`. This class diagram is simplified to better explain the main structure. The data type `Element` contains an XML element and is part of the XCF architecture.

Appendix E

Picture Credits

Credits of foreigen creators are listed below. Images or figures not listed here were created by Sebastian Gieselmann.

Figure 1.1	(middle)	- Image, Copyright: Benedict, Jared C.
	(right)	- Image taken from Burrows, E. (2011) [29]
Figure 2.2		- Figure taken from Mori, Masahiro (1970) [110]
Figure 2.4		- Figure taken from Matthews, G et al. (2000) [102] (p.14)
Figure 2.5		- Figure taken from Matthews, G et al. (2000) [102] (p.165)
Figure 3.2	(middle)	- Image taken from Bazo, Daniel (2010) [7]
	(right)	- Image: Siepmann, Frederic
Figure 4.1		- Image inspired by Hegel, Frank (2009) [69] (p. 29)
Figure 4.2	(left)	- Image, Copyright: Benedict, Jared C.
Figure 4.3	(both)	- Image: Wienke, Johannes
Figure 4.4	(left)	- Image taken from Sakamoto, Daisuke (2007) [137]
	(right)	- Image taken from Burrows, E. (2011) [29]
Figure 4.5	(right)	- Image: Hegel, Frank
Figure 7.11		- Part of the Choregraph GUI (compare [123])
Figure A.14		- List taken from Bartneck, Christoph (2008) [6]
Figure C.1		- List taken from Aschenberner, Bianca (2005) [4]

References

[1] Julie A. Adams and Hande Kaymaz-Keskinpala. "Analysis of Perceived Workload when using a PDA for Mobile Robot Teleoperation". In: *Internatlonal Conference on Robotics and Autornation* 4 (2004), pp. 4128–4133.

[2] Michael Argyle and Janet Dean. "Eye-Contact, Distance and Affiliation". In: *Sociometry* Vol. 28 (1965), pp. 289–304.

[3] Michael Argyle and Roger Ingham. "Gaze, Mutual Gaze, and Proximity". In: *Semiotica* 6.1 (Jan. 1972), pp. 32–49.

[4] Bianca Aschenberner and Christian Weiss. "Phoneme-Viseme Mapping for German Video-Realistic Audio-Visual-Speech-Synthesis". In: *Citeseer* (2005).

[5] F. Gregory Ashby, Vivian V. Valentin, and U. Turken. "The effects of positive affect and arousal on working memory and executive attention - Neurobiology and computational models". In: *Emotional Cognition: From Brain to Behaviour*. Ed. by S. Moore & M. Oaksford. Amsterdam 2002, pp. 245–287.

REFERENCES

[6] Christoph Bartneck, Dana Kulic, and Elizabeth Croft. "Measuring the anthropomorphism , animacy , likeability , perceived intelligence , and perceived safety of robots ." In: *Proceedings of the Metrics for Human-Robot Interaction* (2008), pp. 37–44.

[7] Daniel Bazo et al. "Design and Testing of a Hybrid Expressive Face for a Humanoid Robot". In: *International Conference on Intelligent Robots and Systems*. 2010, pp. 5317–5322.

[8] Nikolaus Bee, Stefan Franke, and Elisabeth Andrea. "Relations between facial display, eye gaze and head tilt: Dominance perception variations of virtual agents". In: *2009 3rd International Conference on Affective Computing and Intelligent Interaction and Workshops*. 2009, pp. 1–7.

[9] Cindy L. Bethel et al. "Psychophysiological experimental design for use in human-robot interaction studies". In: *2007 International Symposium on Collaborative Technologies and Systems*. 2007, pp. 99–105.

[10] Mike Blow et al. "Perception of robot smiles and dimensions for human-robot interaction design". In: *Robot and Human Interactive Communication, ROMAN*. 2006, pp. 469–474.

[11] M. Blumberg and C.D. Pringle. "The missing opportunity in organizational research: Some implications for a theory of work performance". In: *Academy of management Review* 7 (1982), pp. 560–569.

REFERENCES

[12] Wolfram Boucsein. *Electrodermal activity (The Springer Series in Behavioral Psychophysiology and Medicine)*. New York: Springer US, 1992.

[13] Elif Bozkurt et al. "Comparison of phoneme and viseme based acoustic units for speech driven realistic lip animation". In: *Signal Processing and Communications Applications* (2007), pp. 1–4.

[14] R J Brand, D A Baldwin, and L A Ashburn. "Evidence for'motionese': modifications in mothers' infant-directed action". In: *Developmental Science* 5 (2002), pp. 72–83.

[15] Cynthia Breazeal. "Emotion and sociable humanoid robots". In: *International Journal of Human-Computer Studies* 59.1-2 (July 2003), pp. 119–155.

[16] Cynthia Breazeal. "Emotive qualities in lip-synchronized robot speech". In: *Advanced Robotics*. Vol. 17. 2. 2003, pp. 97–113.

[17] Cynthia Breazeal. "Social Interactions in HRI : The Robot View". In: *Systems, Man, and Cybernetics, Part C: Applications and Reviews* 34.2 (2004), pp. 181–186.

[18] Cynthia Breazeal. "Toward sociable robots". In: *Robotics and Autonomous Systems* 42.3 (2003), pp. 167–175.

[19] Cynthia Breazeal, Guy Hoffman, and Andrea Lockerd. "Teaching and working with robots as a collaboration". In: *International Joint Conference on Autonomous Agents and Multiagent Systems*. 2004, pp. 1030–1037.

REFERENCES

[20] Cynthia Breazeal et al. "Effects of Nonverbal Communication on Efficiency and Robustness in Human-Robot Teamwork". In: *Intelligent Robots and Systems* (2005), pp. 708–713.

[21] Cynthia L Breazeal. *Designing Sociable Robots*. Cambridge, Massachusetts: The MIT Press, 2002, p. 263.

[22] Albert van Breemen. "Bringing Robots To Life : Applying Principles Of Animation To Robots". In: *Proceedings of the International Conference for Human-computer Interaction, CHI*. 2004.

[23] Albert van Breemen, Xue Yan, and Bernt Meerbeek. "iCat: an animated user-interface robot with personality". In: *Proceedings of the fourth international joint conference on Autonomous agents and multiagent systems*. 2005, pp. 143–144.

[24] Frank Broz et al. "Towards Automated Human-Robot Mutual Gaze". In: *International Conference on Advances in Computer-Human Interactions, ACHI* (2011), pp. 222–227.

[25] Vicki Bruce. "What the human face tells the human mind: some challenges for the robot-human interface". In: *Proceedings IEEE International Workshop on Robot and Human Communication*. 1992, pp. 44–51.

[26] Steve Burbeck. *Applications Programming in Smalltalk-80 : How to use Model-View-Controller (MVC)*. Mvc. Palo Alto, CA: Softsmarts, Incorporated, 1992, p. 22.

REFERENCES

[27] Robert Burke et al. "Creature smarts: The art and architecture of a virtual brain". In: *Game Developers Conference.* 2001, pp. 147–166.

[28] Felix Burkhardt. "Simulation emotionaler Sprechweise mit Sprachsyntheseverfahren". PhD thesis. Technical University of Berlin, 2000, p. 211.

[29] E. Burrows. "The BIRTh Of a ROBOT RaCe". In: *Engineering & Technology* 6.10 (2011), pp. 46–48.

[30] John T. Cacioppo, Louis G. Tassinary, and Gary G. Berntson. *Handbook of psychophysiology.* Cambridge: Cambridge Univ Pr, 2007.

[31] L G Carney and R M Hill. "The nature of normal blinking patterns". In: *Acta Ophthalmol* 60.3 (1982), pp. 427–433.

[32] Jeanie Chan and Goldie Nejat. "Minimizing Task-Induced Stress in Cognitively Stimulating Activities using an Intelligent Socially Assistive Robot". In: *Ro-Man.* 2011, pp. 296–301.

[33] Nicole Chovil. "Social Determinants of Facial Displays". In: *Journal of Nonverbal Behavior* 15.3 (1991), pp. 141–154.

[34] Herbert H. Clark and Edward F. Schaefer. "Contributing to Discourse". In: *Cognitive Science* 13.2 (Apr. 1989), pp. 259–294.

[35] James Clark. "The design of RELAX NG". In: *http://www.thaiopensource.com/relaxng/design.html.* 2001, pp. 1–11.

249

REFERENCES

[36] Brian L. Connelly et al. "Signaling Theory: A Review and Assessment". In: *Journal of Management* 37.1 (2011), pp. 39–67.

[37] Albert B. Craig. "Heart Rate Responses To Apneic Underwater Diving and To Breath Holding in Man." In: *Journal of applied physiology* 18 (Sept. 1963), pp. 854–62.

[38] Jacob W Crandall and Michael A Goodrich. "Characterizing Efficiency of Human Robot Interaction: A Case Study of Shared-Control Teleoperation". In: *Intlernational Conference on Intelligent Robots and Systems*. 2002, pp. 1–6.

[39] Lee Cronk. "The application of animal signaling theory to human phenomena: some thoughts and clarifications". In: *Social Science Information* 44.4 (Dec. 2005), pp. 603–620.

[40] Daniel Clement Dennett. *The intentional stance*. Cambridge, Massachusetts: MIT Press, 1987, p. 402.

[41] Carl Disalvo and Francine Gemperle. "From Seduction to Fulfillment: The Use of Anthropomorphic Form in Design". In: *Proceedings of the international conference on Designing pleasurable products and interfaces* (2003), pp. 67–72.

[42] Judith Donath. "Mediated Faces". In: *Cognitive Technology: Instruments of Mind* (2001), pp. 373–390.

[43] Brian R. Duffy. "Anthropomorphism and the social robot". In: *Robotics and Autonomous Systems* 42.3-4 (Mar. 2003), pp. 177–190.

REFERENCES

[44] Brian R Duffy et al. "What is a Social Robot?" In: *10th Irish Conference on Artificial Intelligence & Cognitive Science* (1999).

[45] E. Duffy. *Activation and behavior*. New York: John Wiley & Sons Inc, 1962.

[46] Arjan Egges and Nadia Magnenat-Thalmann. "Emotional communicative body animation for multiple characters". In: *First International Workshop on Crowd Simulation (V-Crowds)*. Citeseer. 2005, pp. 31–40.

[47] Arjan Egges and Tom Molet. "Personalised real-time idle motion synthesis". In: *12th Pacific Conference on Computer Graphics and Applications*. Ieee, 2004, pp. 121–130.

[48] Paul Ekman and Wallace V Friesen. "Nonverbal leakage and clues to deception." In: *Psychiatry* 32.1 (Feb. 1969), pp. 88–106.

[49] Nicholas Epley, Adam Waytz, and John T. Cacioppo. "On seeing human: A three-factor theory of anthropomorphism". In: *Psychological Review* 114.4 (2007), pp. 864–886.

[50] Craig Evinger et al. "Not looking while leaping: the linkage of blinking and saccadic gaze shifts". In: *Experimental Brain Research* 100.2 (1994), pp. 337–344.

[51] Friederike Eyssel and Dieta Kuchenbrandt. "Manipulating anthropomorphic inferences about NAO: The role of situational and dispositional aspects of effectance". In: *Ro-Man 2011*. 2011, pp. 467–472.

REFERENCES

[52] Friederike Eyssel et al. "Anthropomorphic inferences from emotional nonverbal cues: A case study". In: *RoMan 2010*. 2010, pp. 646–651.

[53] Terrance Fong, Illah Nourbakhsh, and Kerstin Dautenhahn. "A survey of socially interactive robots". In: *Robotics and Autonomous Systems* 42.3-4 (Mar. 2003), pp. 143–166.

[54] Masahiro Fujita. "AIBO: Toward the Era of Digital Creatures". In: *The International Journal of Robotics Research* 20.10 (Oct. 2001), pp. 781–794.

[55] M Garau et al. "The impact of avatar realism and eye gaze control on perceived quality of communication in a shared immersive virtual environment". In: *Proceedings of the SIGCHI conference on Human factors in computing systems*. ACM. 2003, pp. 529–536.

[56] M Garau et al. "The impact of eye gaze on communication using humanoid avatars". In: *Proceedings of the SIGCHI conference on Human factors in computing systems*. ACM. 2001, p. 316.

[57] Michael J Gielniak and Andrea L Thomaz. "Generating Anticipation in Robot Motion". In: *Ro-Man 2011*. 2011, pp. 449–454.

[58] Sebastian Gieselmann, Marc Hanheide, and Britta Wrede. "Remembering interaction episodes: an unsupervised learning approach for a humanoid robot". In: *Humanoids*. 2010, pp. 566–571.

REFERENCES

[59] Jennifer Goetz, Sara Kiesler, and Aaron Powers. "Matching Robot Appearance and Behavior to Tasks to Improve Human-Robot Cooperation". In: *Ro-Man*. 2003, pp. 55–60.

[60] Barbara Gonsior et al. "Improving Aspects of Empathy and Subjective Performance for HRI through Mirroring Facial Expressions". In: *Ro-Man*. 2011, pp. 350–356.

[61] A. Gonzalez et al. "Sound quality of low-frequency and car engine noises after active noise control". In: *Journal of Sound and Vibration* 265.3 (Aug. 2003), pp. 663–679.

[62] David Gouaillier et al. "Mechatronic design of NAO humanoid". In: *2009 IEEE International Conference on Robotics and Automation* (May 2009), pp. 769–774.

[63] S.E. Guthrie. "Anthropomorphism: a definition and a theory". In: *Anthropomorphism, anecdotes, and animals*. 1997, pp. 50–58.

[64] Marc Hanheide et al. "Who am I talking with? A Face Memory for Social Robots". In: *International Conference on Robotics and Automation*. IEEE. Pasadena, CA, USA 2008, pp. 3660–3665.

[65] David Hanson. "Exploring the Aesthetic Range for Humanoid Robots". In: *Proceedings of the ICCS/CogSci-2006 long symposium: Toward social mechanisms of android science*. 2006, pp. 39–42.

[66] Markus Häring, Nikolaus Bee, and Elisabeth Andre. *Creation and Evaluation of Emotion Expression with Body Movement, Sound and Eye Color for Humanoid Robots*. 2011, pp. 204–209.

REFERENCES

[67] Jennifer A. Healey and RRosalind W. Picard. "Detecting Stress During Real-World Driving Tasks Using Physiological Sensors". In: *IEEE Transactions on Intelligent Transportation Systems* 6.2 (June 2005), pp. 156–166.

[68] M Heerink et al. "Studying the acceptance of a robotic agent by elderly users". In: *International Journal of Assistive Robotics and Mechatronics* 7.3 (2006), pp. 33–43.

[69] Frank Hegel. *Gestalterisch konstruktiver Entwurf eines sozialen Roboters*. Uelvesbüll: Der andere Verlag, 2010, p. 222.

[70] Frank Hegel, Friederike Eyssel, and Britta Wrede. "The Social Robot Flobi : Key Concepts of Industrial Design". In: *International Symposium on Robot and Human Interactive Communication*. 2010, pp. 107–112.

[71] Frank Hegel et al. "Playing a different imitation game: Interaction with an Empathic Android Robot". In: *International Conference on Humanoid Robots*. Ieee, 2006, pp. 56–61.

[72] Frank Hegel et al. "Towards a Typology of Meaningful Signals and Cues in Social Robotics". In: *Ro-Man*. 2011, pp. 72–78.

[73] Javier Hernandez, Rob R Morris, and Rosalind W Picard. "Call Center Stress Recognition with Person-Specific Models". In: *Affective Computing and Intelligent Interaction*. 2011, pp. 125–134.

REFERENCES

[74] Pamela J. Hinds, Teresa L. Roberts, and Hank Jones. "Whose job is it anyway? A study of human-robot interaction in a collaborative task". In: *Human-Computer Interaction* 19.1 (2004), pp. 151–181.

[75] Laura Hoffmann et al. "Media Equation revisited . Do users show polite reactions towards an embodied agent ?" In: *Intelligent Virtual Agents*. 2009, pp. 159–165.

[76] Chien-ming Huang and Bilge Mutlu. "Robot Behavior Toolkit : Generating Effective Social Behaviors for Robots". In: *HRI*. 2012, pp. 25–32.

[77] Kazuko Itoh et al. "Development of a Bioinstrumentation System in the Interaction between a Human and a Robot". In: *International Conference on Intelligent Robots and Systems*. Ieee, 2006, pp. 2620–2625.

[78] Laurent Itti, Christof Koch, and Ernst Niebur. "A model of saliency-based visual attention for rapid scene analysis". In: *IEEE Transactions on pattern analysis and machine intelligence* 20.11 (1998), pp. 1254–1259.

[79] O.P. John, EM Donahue, and R. Kentle. "Big five inventory". In: *Berkeley, CA: Berkeley Institute of Personality and Social Research, University of California* (1991).

[80] Takayuki Kanda, Michihiro Shimada, and Satoshi Koizumi. "Children Learning with a Social Robot". In: *HRI*. 2012, pp. 351–358.

REFERENCES

[81] Frederic Kaplan. "Who is afraid of the humanoid? Investigating cultural differences in the acceptance of robots". In: *International journal of humanoid robotics* 1.3 (2004), pp. 1–16.

[82] Klaus Kemp. *Less and More: The Design Ethos of Dieter Rams*. Die Gestalten Verlag, 2010, p. 808.

[83] Cory D Kidd and Cynthia Breazeal. "Effect of a Robot on User Perceptions". In: *Intelligent Robots and Systems, IROS*. 2004, pp. 3559–3564.

[84] Sara Kiesler and Jennifer Goetz. "Mental models of robotic assistants". In: *CHI'02 extended abstracts on Human factors in computing systems*. 2002, pp. 576–577.

[85] Sören Krach et al. "Can machines think? Interaction and perspective taking with robots investigated via fMRI". In: *PLoS ONE* 3.7 (2008).

[86] Alfred Kranstedt, Stefan Kopp, and Ipke Wachsmuth. "MURML: A multimodal utterance representation markup language for conversational agents". In: *Proc. of the AAMAS Workshop on Embodied conversational agents–Lets specify and evaluate them*. 2002.

[87] Dana Kulic and Elizabeth Croft. "Anxiety detection during human-robot interaction". In: *2005 IEEE/RSJ International Conference on Intelligent Robots and Systems*. Ieee, 2005, pp. 616–621.

REFERENCES

[88] Yoshihiro Kuroki et al. "Motion Creating System for A Small Biped Entertainment Robot". In: *Intlernational Conference on Intelligent Robots and Systems* 2 (2003), pp. 1394–1399.

[89] Fabien D. Legrand et al. "Interactive-Virtual Reality (IVR) Exercise: An Examination of In-Task and Pre-to-Post Exercise Affective Changes". In: *Journal of Applied Sport Psychology* 23.1 (Jan. 2011), pp. 65–75.

[90] Changchun Liu, Pramila Rani, and Nilanjan Sarkar. "Affective State Recognition and Adaptation in Human-Robot Interaction: A Design Approach". In: *2006 IEEE/RSJ International Conference on Intelligent Robots and Systems.* Ieee, 2006, pp. 3099–3106.

[91] Frank Loemker. *IceWing.* 2007. URL: `http : / / icewing . sourceforge.net`.

[92] Katrin S. Lohan et al. "Tutor Spotter: Proposing a Feature Set and Evaluating It in a Robotic System". In: *International Journal of Social Robotics* 4.2 (Dec. 2011), pp. 131–146.

[93] Katrin Solveig Lohan et al. "Does embodiment affect tutoring behavior ?" In: *International Conference on Development and Learning, ICDL.* 2010.

[94] Manja Lohse. "The role of expectations in HRI". In: *New Frontiers in Human-Robot Interaction.* 2009.

[95] Manja Lohse et al. "What can I do for you? Appearance and Application of Robots". In: *Proceedings of AISB* 7 (2007), pp. 121–126.

REFERENCES

[96] Manuel Lopes et al. "Biomimetic Eye-Neck Coordination". In: *IEEE 8th International Conference on Development and Learning*. 2009, pp. 1–8.

[97] Holger Luczak, Matthias Roetting, and Ludger Schmidt. "Let's talk: anthropomorphization as means to cope with stress of interacting with technical devices." In: *Ergonomics* 46.13-14 (2003), pp. 1361–74.

[98] Ingo Lütkebohle et al. "The Bielefeld Anthropomorphic Robot Head "Flobi"". In: *2010 IEEE International Conference on Robotics and Automation*. IEEE. Anchorage, Alaska: IEEE, 2010, pp. 3384–3391.

[99] Mark Maat, Khiet P Truong, and Dirk Heylen. "How turn-taking strategies influence users impressions of an agent". In: *Intelligent Virtual Agents*. 2010, pp. 441–453.

[100] Julie C Main et al. "Integrating gaze direction and sexual dimorphism of face shape when perceiving the dominance of others". In: *Perception* 38.9 (2009), pp. 1275–1283.

[101] Eliseo Stefano Maini, Luigi Manfredi, and Cecilia Laschi. "Bioinspired velocity control of fast gaze shifts on a robotic anthropomorphic head". In: *Autonomous Robots*. Vol. 25. 1. 2008, pp. 37–58.

[102] G Matthews et al. *Human Performance: Cognition, Stress, and Individual Differences*. Hove, East Sussex: Psychology Press Ltd, 2000, p. 416.

REFERENCES

[103] Gerald Matthews et al. "Emotional intelligence, personality, and task-induced stress." In: *Journal of experimental psychology. Applied* 12.2 (June 2006), pp. 96–107.

[104] John Maynard-Smith. *Animal Signals*. Oxford: Oxford University Press, 2003, p. 176.

[105] Giorgio Metta, Paul Fitzpatrick, and Lorenzo Natale. "YARP: Yet Another Robot Platform". In: *International Journal of Advanced Robotic Systems* 3.1 (2006), p. 1.

[106] Giorgio Metta et al. "The iCub humanoid robot: an open platform for research in embodied cognition". In: *PerMIS: Performance Metrics for Intelligent Systems Workshop. Aug.* 2008, pp. 19–21.

[107] T Minato et al. "Does gaze reveal the human likeness of an android?" In: *The 4th International Conference on Development and Learning, 2005. Proceedings*. 2005, pp. 106–111.

[108] T Minato et al. "Evaluating the human likeness of an android by comparing gaze behaviors elicited by the android and a person". In: *Advanced robotics: the international journal of the Robotics Society of Japan* 20.10 (2006), p. 1147.

[109] Takashi Minato et al. "Development of an Android Robot for Studying Human-Robot Interaction". In: *Innovations in Applied Artificial Intelligence*. 2004, pp. 424–434.

[110] Masahiro Mori. "The Uncanny Valley". In: *Energy*. Vol. 7. 4. 1970, pp. 33–35.

REFERENCES

[111] Helmut Morsbach. "Aspects of nonverbal communication in Japan". In: *The Journal of Nervous and Mental Disease* 157.4 (1973), p. 262.

[112] B Mutlu, J Forlizzi, and J Hodgins. "A storytelling robot: Modeling and evaluation of human-like gaze behavior". In: *International Conference on Humanoid Robots*. 2006, pp. 518–523.

[113] Bilge Mutlu. "Designing Gaze Behavior for Humanlike Robots". PhD thesis. Carnegie Mellon University, 2009, p. 249.

[114] Bilge Mutlu et al. "Nonverbal Leakage in Robots : Communication of Intentions through Seemingly Unintentional Behavior". In: *Proceedings of the 4th ACM/IEEE international conference on Human robot interaction* 2.1 (2009), pp. 69–76.

[115] Clifford Nass and Youngme Moon. "Machines and Mindlessness: Social Responses to Computers". In: *Journal of Social Issues* 56.1 (Jan. 2000), pp. 81–103.

[116] Clifford Nass et al. "Can computer personalities be human personalities?" In: *Conference companion on Human factors in computing systems* 43 (1995), pp. 223–239.

[117] Winfried Nöth. *Handbook of Semiotics*. Bloomington, IN: Indiana Univ Pr, 1995, p. 576.

[118] Kristine L. Nowak and Frank Biocca. "The Effect of the Agency and Anthropomorphism on Users' Sense of Telepresence, Copresence, and Social Presence in Virtual Environments". In: *Presence: Teleoperators and Virtual Environments* 12.5 (Oct. 2003), pp. 481–494.

REFERENCES

[119] Salvatore Parise et al. "Cooperating with Life-like Interface Agents". In: *Computers in Human Behavior* 15 (1998), pp. 123–142.

[120] Annika. Peters et al. "Hey robot, get out of my way." In: *Behavior Monitoring and Interpretation. Well-Being. Vol 9. Ambient Intelligence and Smart Environments*. Ed. by Björn Gottfried and Hamid K. Aghajan. IOS Press, 2011.

[121] Karola Pitsch et al. "On the loop of action modification and the recipient s gaze in". In: *Gesture and Speech in Interaction* 24 (2009), pp. 24–26.

[122] M. Poel et al. "Gaze behaviour, believability, likability and the iCat". In: *Ai & Society* 24.1 (Feb. 2009), pp. 61–73.

[123] E Pot et al. "Choregraphe: a Graphical Tool for Humanoid Robot Programming". In: *Robot and Human Interactive Communication (Ro-Man)*. 2009, pp. 46–51.

[124] Morgan Quigley et al. "ROS : an open-source Robot Operating System". In: *ICRA Workshop on Open Source Software*. Vol. 3. 2009.

[125] Pramila Rani et al. "Anxiety detecting robotic system towards implicit human-robot collaboration". In: *Robotica* 22.1 (Jan. 2004), pp. 85–95.

[126] Pramila Rani et al. "Online Stress Detection using Psychophysiological Signal for Implicit Human-Robot Cooperation". In: *Robotica* 20.6 (2002), pp. 673–686.

REFERENCES

[127] Byron Reeves and Clifford Nass. *The Media Equation: How People Treat Computers, Television, and New Media like Real People and Places (CSLI Lecture Notes)*. Cambridge: Cambridge University Press, 1996, p. 323.

[128] Laurel D. Riek, Philip C. Paul, and Peter Robinson. "When my robot smiles at me: Enabling human-robot rapport via real-time head gesture mimicry". In: *Journal on Multimodal User Interfaces* 3.1-2 (Nov. 2009), pp. 99–108.

[129] Laurel D. Riek et al. "How anthropomorphism affects empathy toward robots". In: *Proceedings of the 4th ACM/IEEE international conference on Human robot interaction - HRI*. 2009, pp. 245–246.

[130] Nina Riether et al. "Social facilitation with social robots?" In: *Proceedings of the seventh annual ACM/IEEE international conference on Human-Robot Interaction - HRI*. 2012, pp. 41–47.

[131] Christina Röcke and Daniel Grühn. "German Translation of the PANAS-X". In: *Unpublished manuscript, Free University Berlin* (2003).

[132] K J Rohlfing et al. "How can multimodal cues from child-directed interaction reduce learning complexity in robots?" In: *Advanced Robotics* 20 (2006), pp. 1183–1199.

[133] J.A. Russell and A. Mehrabian. "Evidence for a three-factor theory of emotions". In: *Journal of research in Personality* 11 (1977), pp. 273–294.

REFERENCES

[134] James A. Russell and Lisa Feldmann Barrett. "Core affect, prototypical emotional episodes, and other things called emotion: dissecting the elephant." In: *Journal of personality and social psychology* 76.5 (May 1999), pp. 805–19.

[135] Martin Saerbeck and Albert J N Van Breemen. "Design guidelines and tools for creating believable motion for personal robots". In: *Robot and Human Interactive Communication (Ro-Man)*. 2007, pp. 386–391.

[136] Tomoko Saito et al. "Relationship between Interaction with the Mental Commit Robot and Change of Stress Reaction of the Elderly". In: *Computational Intelligence in Robotics and Automation*. 2003, pp. 119–124.

[137] Daisuke Sakamoto et al. "Android as a telecommunication medium with a human-like presence". In: *Proceeding of the ACM/IEEE international conference on Human-robot interaction - HRI*. 2007, pp. 193–200.

[138] Nilanjan Sarkar. "Psychophysiological Control Architecture for Human-Robot Coordination - Concepts and Initial Experiments". In: *International Conference on Robotics & Automation*. 2002, pp. 3719–3724.

[139] B.J. Scholl and P.D. Tremoulet. "Perceptual causality and animacy". In: *Trends in cognitive sciences* 4.8 (2000), pp. 299–309.

REFERENCES

[140] Marc Schröder and Jürgen Trouvain. "The German Text-to-Speech Synthesis System MARY : A Tool for Research ," in: *International Journal of Speech Technology* 6 (2003), pp. 365–377. URL: `http://mary.dfki.de/`.

[141] Cornelia Setz et al. "Discriminating Stress From Cognitive Load Using a Wearable EDA Device". In: *Information Technology In Biomedicine* 14.2 (2010), pp. 410–417.

[142] J. Seyama and R.S. Nagayama. "The Uncanny Valley : Effect of Realism on the Impression of Artificial Human Faces". In: *Presence: Teleoperators and Virtual Environments* 16.4 (2007), pp. 337–351.

[143] Yuan Shi et al. "Personalized Stress Detection from Physiological Measurements". In: *International Symposium on Quality of Life Technology*. 2010.

[144] Michihiro Shimada and Hiroshi Ishiguro. "Motion Behavior and its Influences on Human-likeness in an Android Robot". In: *30th Annual meeting of Cognitive Science Society*. 2008.

[145] Elaine Short et al. "No fair!!: an interaction with a cheating robot". In: *International conference on Human-robot interaction*. ACM, 2010, pp. 219–226.

[146] Hyunsoo Song et al. "Design of idle motions for service robot via video ethnography". In: *he 18th IEEE International Symposium on Robot and Human Interactive Communication, Ro-Man*. 2009, pp. 195–199.

REFERENCES

[147] Michael Spence. "Job Market Signaling". In: *The quarterly journal of Economics* 87.3 (1973), pp. 355–374.

[148] Frank Thomas and Ollie Johnston. *The illusion of life: Disney animation*. New York: Hyperion, 1995, p. 576.

[149] Hannes Vilhjalmsson et al. "The Behavior Markup Language : Recent Developments and Challenges". In: *Intelligent Virtual Agents*. 1. 2007, pp. 99–111.

[150] Paul Viola and Michael Jones. "Rapid object detection using a boosted cascade of simple features". In: *Proceedings of the IEEE Computer Society Conference on Computer Vision and Pattern Recognition. CVPR*. 2001, pp. I–511–I–518.

[151] Michiel Visser, Mannes Poel, and Anton Nijholt. "Classifying visemes for automatic lipreading". In: *Text, Speech and Dialogue*. Springer, 1999, pp. 843–843.

[152] Anna-lisa Vollmer. "Measurement and Analysis of Interactive Behavior in Tutoring Action with Children and Robots". PhD thesis. Universität Bielefeld, 2011, p. 143.

[153] Kazuyoshi Wada, Takanori Shibata, and Tomoko Saito. "Effects of Robot-Assisted Activity for Elderly People and Nurses at a Day Service Center". In: *Therapy* 92.11 (2004).

[154] Kazuyoshi Wada et al. "Psychological and Social Effects of One Year Robot Assisted Activity on Elderly People at a Health Service Fa-

REFERENCES

cility for the Aged". In: *Proceedings of the 2005 IEEE International Conference on Robotics and Automation.* 2005, pp. 2785–2790.

[155] Rebecca M. Warner and David B. Sugarman. "Attributions of personality based on physical appearance, speech, and handwriting." In: *Journal of Personality and Social Psychology* 50.4 (1986), p. 792.

[156] Keith Waters and Tom Levergood. "An automatic lip-synchronization algorithm for synthetic faces". In: *MULTIMEDIA '94: Proceedings of the second ACM international conference on Multimedia.* 1994, pp. 149–156.

[157] Keith Waters and Tom Levergood. "DECface: A system for synthetic face applications". In: *Multimedia Tools and Applications* 1.4 (Nov. 1995), pp. 349–366.

[158] David Watson and Lee Anna Clark. "The PANAS-X : Manual for the Positive and Negative Affect Schedule - Expanded Form". In: *Psychology* (1999).

[159] David Watson, Lee Anna Clark, and Auke Tellegen. "Development and validation of brief measures of positive and negative affect: the PANAS scales." In: *Journal of personality and social psychology* 54.6 (1988), pp. 1063–1070.

[160] Paul Watzlawick, Janet Helmick Beavin, and M D Don D Jackson. *Pragmatics of Human Communication: A Study of Interactional Patterns, Pathologies, and Paradoxes.* New York: Norton, 1967, p. 296.

REFERENCES

[161] A. Wells and G. Matthews. *Attention and emotion: A clinical perspective*. Hove, UK: Lawrence Erlbaum Associates, Inc, 1994, p. 402.

[162] J.C. Wells and Others. "SAMPA computer readable phonetic alphabet". In: *Handbook of Standards and Resources for Spoken Language Systems*. Ed. by R. Gibbon, D., Moore, R. and Winski. Vol. 4. Berlin and New York 1997.

[163] Richard Williams. *The Animator's Survival Kit*. London: Faber & Faber, 2001, p. 352.

[164] Sebastian Wrede. "An Information-Driven Architecture for Cognitive Systems Research". PhD thesis. Bielefeld, Germany: Technical Faculty – Bielefeld University, 2008.

[165] Robert M. Yerkes and John D. Dodson. "The relation of strength of stimulus to rapidity of habit-formation". In: *Journal of comparative neurology and psychology* 18.5 (1908), pp. 459–482.

[166] Amotz Zahavi. "Mate selection-a selection for a handicap." In: *Journal of theoretical biology* 53.1 (Sept. 1975), pp. 205–14.

[167] RB Zajonc. "Social facilitation". In: *Science* 149.3681 (1965), pp. 269–274.

[168] L.A. Zebrowitz. *Reading faces: Window to the soul?* Boulder, Colorado: Westview Press, 1997, p. 288.

REFERENCES

[169] Jan Zwickel and Hermann J. Müller. "Eye Movements as a Means to Evaluate and Improve Robots". In: *International Journal of Social Robotics* 1.4 (Oct. 2009), pp. 357–366.

Der disserta Verlag bietet die kostenlose Publikation
Ihrer Dissertation als hochwertige
Hardcover- oder Paperback-Ausgabe.

Fachautoren bietet der disserta Verlag
die kostenlose Veröffentlichung professioneller Fachbücher.

Der disserta Verlag ist Partner für die Veröffentlichung
von Schriftenreihen aus Hochschule und Wissenschaft.

Weitere Informationen auf www.disserta-verlag.de